Advanced Fossil Energy Utilization 2013

Topical Conference at the 2013 AIChE Annual Meeting: Global Challenges for Engineering a Sustainable Future

San Francisco, California, USA
3-8 November 2013

ISBN: 978-1-63439-031-6

Printed from e-media with permission by:

Curran Associates, Inc.
57 Morehouse Lane
Red Hook, NY 12571

Printed by Curran Associates, Inc. (2014)

AIChE
3 Park Avenue
New York, NY 10016-5991

Phone: (203) 702-7660
Fax: (203) 775-5177

www.aiche.org

Additional copies of this publication are available from:

Curran Associates, Inc.
57 Morehouse Lane
Red Hook, NY 12571 USA
Phone: 845-758-0400
Fax: 845-758-2634
Email: curran@proceedings.com
Web: www.proceedings.com

TABLE OF CONTENTS

(45a) Computational Carbon Capture

Smit, B., University of Amsterdam

Originally presented on: 11/4/2013 8:30:00 - 8:55:00

In this presentation we describe how computational techniques can be used to screen materials for carbon capture. Our starting point is the question how to define the optimal material. We introduce the concept of parasitic energy as a metric to compare different materials; the best material is the material that minimizes the loss of efficiency of a power plant. To compute this parasitic energy one need information on the mixture isotherms of the various components of flue gasses.[1]

We show how quantum chemical calculations can be used to predict the adsorption energies. In addition, we show how these calculations can be used to develop a force field that allows us to predict the adsorption properties.[2] We will apply these methods for materials for which experimental data is lacking, or for materials that have not yet been synthesized.

1 L.-C. Lin, A. H. Berger, R. L. Martin, J. Kim, J. A. Swisher, K. Jariwala, C. H. Rycroft, A. S. Bhown, M. W. Deem, M. Haranczyk, and B. Smit, *In silico screening of carbon-capture materials* Nat Mater **11** (7), 633 (2012) http://dx.doi.org/10.1038/nmat3336

2 A. Dzubak, L.-C. Lin, J. Kim, J. A. Swisher, R. Poloni, S. N. Maximoff, B. Smit, and L. Gagliardi, *Ab-initio Carbon Capture in Open-Site Metal Organic Frameworks* Nat Chem **4**, 810 (2012) http://dx.doi.org/0.1038/NCHEM.1432

(45b) Polyethyleneimine-Functionalized Polyamide-Imide Hollow Fiber Sorbents for Post-Combustion CO2 Capture

Li, F. S., Georgia Institute of Technology
Qiu, W., Georgia Institute of Technology
Lively, R. P., Georgia Institute of Technology
Lee, J. S., Georgia Institute of Technology
Rownaghi, A. A., Georgia Institute of Technology
Koros, W. J., Georgia Institute of Technology
Originally presented on: 11/4/2013 8:55:00 - 9:13:00

Carbon dioxide emitted from existing coal-fired power plants causes a major environmental concern due to possible linkages to global climate change. Therefore developing materials for capturing carbon dioxide has become an important issue for this application. In this study, we expand previous work focused on aminosilane-functionalized polymeric hollow fiber sorbents by introducing a new class of polyethyleneimine (PEI) functionalized polymeric hollow fiber sorbents for post-combustion carbon dioxide capture. The study proves that the open-walled porous morphology of the polymeric hollow fibers can be maintained after functionalizing with PEI. Different molecular weight PEIs were studied as modification reagents on polyamide-imide hollow fibers. This imide ring-opening modification introduces two amide functional groups and was confirmed via FTIR-ATR. The carbon dioxide equilibrium sorption capacities of PEI functionalized polyamide-imide materials were characterized using both pressure decay and gravimetric sorption methods. For equivalent PEI concentrations, polyamide-imide fibers functionalized with lower molecular weight PEI exhibited higher sorption capacities. Our hypothesis to explain this observation is that the imide ring opening event is substantially more prevalent in the low PEI molecular weight samples. Diffusional and steric limitations become significant hindrances for efficient tethering in the high PEI molecular weight samples. The effect of water in the ring-opening reaction was also studied. Up to a critical value, water in the reaction mixture enhanced the degree of functionalization. The optimal water concentration found in this reaction is 10% (w/w).

(45c) Surface Functionalization Of Mesoporous Sorbents For Post-Combustion CO2 Capture

He, J., Stanford University
To, J., Stanford University
Lyons, C. T., Stanford University
Haghpanah, R., Stanford University
Rupp, E. C., National Energy Technology Laboratory
Bao, Z., Stanford University
Stack, D., Stanford University
Wilcox, J., Stanford University
Originally presented on: 11/4/2013 9:13:00 - 9:31:00

Solid sorbent technologies for CO_2 capture have several advantages over the traditional amine-based solvent absorption approaches. For instance, within an adsorption-based approach water is absent, which decreases the energy requirements associated with regeneration since heating a large amount of water is the greatest energetic expense associated with CO_2 capture using solvent-based approaches. Another benefit of using solid sorbents is the ability to tune the structural parameters of the material, as well as perform chemical modifications to enhance CO_2uptake. In addition, materials such as carbon possess favorable heat conduction properties.

Mesoporous silica materials have been highly developed in materials chemistry, and allow for more extensive chemical modification for CO_2capture. While developing chemistry amenable to both silica and carbon surfaces, mesoporous carbon-based framework development will allow for heat to be dissipated readily during the adsorption process, which will lead to maximum capacity, in addition to the ease of heat transfer into the system for regeneration.

We aim to use careful chemical modification of mesoporous silica- and carbon-based materials to achieve highly selective, recyclable, and efficient materials for CO_2 adsorption. Through careful tuning of the surface CO_2 adsorption chemistry, we can control the thermodynamics and kinetics of carbon dioxide adsorption to allow for efficient material development. Our inspiration for this idea comes from carbonic anhydrase (CA), an enzyme that can selectively adsorb and regenerate CO_2. CAs are found in the red blood cells of mammals, and are used to capture CO_2 as bicarbonate, which can easily dissolve in blood and transported to the lungs. The enzyme active site has been tuned such that the overall transformation is both kinetically fast and near thermodynamically neutral, allowing for both rapid and reversible CO_2 conversion. It is our aim to apply this design principle to sorbent-based technologies to enhance the mass transfer of CO_2 from the gas to adsorbed phase, in addition to enhancing the kinetics associated with adsorption and desorption (regeneration) processes.

(45d) Capture of Carbon Dioxide With Tetraethylenepentamine-Functionalized Industrial Carbon Nanotubes

Liu, Q.,
Zheng, S.,
He, Y.,
Shi, Y.,
Originally presented on: 11/4/2013 9:31:00 - 9:49:00

Industrial carbon nanotubes (I-CNTs) is a low-cost substitute for commercially available multi-walled carbon nanotubes (CCNTs). In this work, I-CNTs were functionalized with tetraethylenepentamine (TEPA) for CO_2 capture. The performance of I-CNTs-TEPA was compared with that of CCNTs. It was found that I-CNTs-TEPA exhibited high CO_2 adsorption capacity similar to CCNTs-TEPA. The adsorption capacity of I-CNTs-TEPA was in the range of 2.145 to 3.088 mmol/g depending on adsorption temperatures. Both adsorption isotherms of I-CNTs-TEPA and the isosteric heat of different adsorption capacity were obtained from experiments. H_2O, SO_2 or NO has been found to have no significant impacts on the adsorption capacity of CO_2. I-CNTs-TEPA was further extruded to form granules (I-CNTs-TEPA-G) with a specific mold. The adsorption capacity of I-CNTs-TEPA and I-CNTs-TEPA-G remained almost the same after 5 cycles of adsorption/regeneration with adsorption conditions of 10 vol % CO_2, 500 ppm SO_2, 500 ppm NO, 100% relative humidity, and a temperature of 313 K. Having advantages such as low-cost, easiness in processing, high adsorption capacity, good poison resistance, and decent performance after recycling, I-CNTs-TEPA seems to be a promising adsorbent for CO_2 capture from flue gas.

(45e) Post-Combustion CO2 Capture Using PAI Polymer/Silica/PEI Hollow Fiber Sorbents

Labreche, Y., Georgia Tech
Fan, Y., Georgia Tech
Rezaei, F., Georgia Institute of Technology
Lively, R. P., Georgia Institute of Technology
Jones, C. W., Georgia Institute of Technology
Koros, W. J., Georgia Institute of Technology
Originally presented on: 11/4/2013 9:49:00 - 10:07:00

Post-combustion CO_2 capture using PAI polymer/silica/PEI hollow fiber sorbents

Ying Labreche[1], Yanfang Fang[1], Fateme Rezaei[1], Ryan P. Lively[2], Christopher W. Jones[1,*], William J. Koros[1,*]

[1] Georgia Institute of Technology, 311 Ferst Drive NW, Atlanta, GA 30332-0100, USA

[2] Algenol Biofuels, 28100 Bonita Grande Drive, Bonita Springs, Florida 34315, United States

Abstract

Amine-loaded cellulose acetate (CA) polymer/silica hollow fiber sorbents for post-combustion CO_2 recovery were created successfully by utilizing a post-spinning amine infusion technique [1]. After initial work focusing on CA fibers [1-2], poly(amide-imide) (PAI), fibers were created, as this polymer is more thermally and mechanically stable than CA. Poly(ethyleneimine) (PEI) was infused into PAI/mesoporous silica hollow fiber sorbents during the solvent exchange steps after fiber spinning. After infusing the 50 wt % silica hollow fiber sorbents with 10% PEI solutions in methanol, the fiber sorbents were demonstrated to have an equilibrium CO_2 uptake of ca. 1.1 mmol/g-fiber in TGA tests using dry, 10% CO_2 simulated gas streams.

Several amine-loaded fibers were assembled into shell-and-tube modules and exposed on the shell side at 1 atm and 35 °C to simulated flue gas with an inert tracer (14 mol% CO_2, 72 mol% N_2 and 14 mol% He at 100% R.H.). The humid breakthrough CO_2 capacity and the humid CO_2 and "equilibrium" uptake at 1 atm and 35 °C were tested. The results indicate that the post-spinning infusion method provides a versatile platform for ultimate scale-up of practical amine sorbents for flue gas CO_2 recovery applications.

[1]Y. Labreche, R.P. Lively, F. Rezaei, G. Chen, C.W. Jones, W.J. Koros, Post-spinning infusion of poly(ethyleneimine) into polymer/silica hollow fiber sorbents for carbon dioxide capture *Chem. Eng. J.,* 219 (2013) 166-175

[2] F. Rezaei, R. Lively, Y. Labreche, G. Chen, Y. Fan, W.J. Koros, C.W. Jones, Aminosilane-grafted polymer/silica hollow fiber adsorbents for CO_2 capture from flue gas, *ACS Appl. Mater. Interfaces*, 5 (2013) 3921-3931

(45f) Mg(OH)2-Based Sorbents for Combined CO2 Capture and Storage in Gasification Systems

Fricker, K.,
Park, A. H. A., Columbia University
Originally presented on: 11/4/2013 10:07:00 - 10:25:00

Mg-bearing sorbents, specifically solid or aqueous $Mg(OH)_2$, derived from silicate minerals or industrial wastes, can directly capture and store CO_2 while eliminating an energy intensive sorbent regeneration step. By integrating in-situ mineral carbonation into various energy conversion systems, greater sustainability can be achieved in the use of carboneous fuels. In the water-gas shift reaction, used industrially to tailor synthesis gas composition for downstream applications, gas-solid or slurry phase $Mg(OH)_2$ carbonation will enhance H_2 production as the CO_2 is converted to $MgCO_3$ and the equilibrium is shifted. Steam, already a reactant in the water-gas shift, functions as both a H_2 source and carbonation promoter. This work investigates the pathways of enhanced $Mg(OH)_2$ carbonation at elevated temperatures and CO_2 pressures (up to 400 °C and 15 atm) in the presence of steam as well as in the slurry phase. The extent of carbonation increases dramatically with increasing H_2O loading. Fundamental mechanisms for water enhanced carbonation are suggested, and the reaction conditions responsible for hydrated and anhydrous carbonate product phases are evaluated. The results suggested that a hydrated environment facilitates the formation of intermediate hydrated magnesium carbonate species, which subsequently decompose to $MgCO_3$. The carbonation reaction is integrated with the water gas shift reaction in a gas-solid, flow-through reactor and semi-batch slurry reactor to investigate the effect of in-situ carbonation on water-gas shift reaction characteristics, namely enhanced H_2 yield. Optimal reactor arrangements are proposed. Comprehensive solid analyses via TGA, XRD, UV-Raman, and SEM allow for qualitative and quantitative compositional characterizations of reacted solids and micro GC provides online gas phase analysis.

(45g) A Low Cost, High Capacity Regenerable Sorbent for Post Combustion CO2 Capture

Jayaraman, A., TDA Research, Inc.
Alptekin, G., TDA Research Inc.
Dietz, S., TDA Research Inc.
Copeland, R., TDA Research, Inc.
Rao, A.,

Originally presented on: 11/4/2013 10:25:00 - 10:43:00

TDA Research, Inc (TDA) is developing a low cost, high capacity CO_2 adsorbent and demonstrating its technical and economic viability for post-combustion CO_2 capture for existing pulverized coal-fired power plants under a 3-year DOE funded project. TDA uses an advanced physical adsorbent to selectively remove CO_2 from the flue gas. The sorbent exhibits a much higher affinity for CO_2 than N_2, H_2O or O_2. The sorbent binds CO_2 more strongly than common adsorbents, providing the chemical potential needed to remove the CO_2. However, because CO_2 does not form a true covalent bond with the surface sites, the regeneration can be carried out with only a very small energy input (4.9 kcal per mol of CO_2) compared to amines (14.2 kcal/mol CO_2). We demonstrated the sorbent's efficacy through multiple cycles for removing CO_2 from coal derived flue gas and the sorbent achieved a stable capacity. The presence of acid gases such as SO_2 and NOx (80 and 200 ppmv, respectively) and water vapor up to 15% vol. caused no adverse effect on the CO_2 capacity. The regeneration of the sorbent and the recovery of CO_2 and its pressurization can be achieved by several approaches, including temperature and vacuum swing. Our preliminary analysis showed that TDA's CO_2 capture process will reduce plant efficiency only by 21.7% (compared to 30% reduction for amine scrubbers) while providing 90% capture. A detailed techno-economic analysis using Aspen Plus^{TM} software is being conducted in collaboration with University of California Irvine to estimate the impact of the TDA's CO_2 capture system on plant efficiency and cost of electricity. We will present the results from our Year 1 and Year 2 work at the AIChE meeting.

(45h) Effect of Structural Changes During Partial Calcination of Dolomite-Based Sorbent On CO2 Capture in IGCC

Zarghami, S., Illinois Institute of Technology
Abbasian, J., Illinois Institute of Technology
Originally presented on: 11/4/2013 10:43:00 - 11:01:00

Carbon dioxide emission from fossil fuel combustion and its impact on global warming is one of the most critical environmental issues nowadays. Coal as a main source of produce energy is the most CO_2-intensive fossil fuel. Advanced power generation processes that use gasification technology, such as Integrated Gasification Combined Cycle (IGCC), which offer higher efficiency, are among the leading contenders for power generation in the 21st century. In an IGCC process, because of high pressure, carbon dioxide in the fuel gas is at higher concentration, which can be captured and sequestered at lower costs. Utilization of regenerable MgO-based sorbents has been shown to be an effective method for capturing CO_2 from gasification-based processes at high pressures and temperatures. Dolomite, as an abundant mineral is a ideal source of magnesium for the development of MgO-based sorbents for CO_2 capture. Although the sorbent can be regenerated, the cyclic process is hindered by a slow and incomplete regeneration step resulting in decreased CO_2 capture capacity in the cyclic process.

In an effort to develop effective solutions to improve regenerability of dolomite-based sorbents, as a first step the kinetics of partial calcinations of dolomite was studied. A variable grain size model was developed to predict the rate and the extent of decomposition of dolomite. The size distribution of grains was obtained by measuring the pore size distribution and total porosity of the dolomite particles. The model is able to account for changes in the rate and the extent of reaction and provides an excellent fit to the experimental data.

(111a) Process Configuration Studies of Conventional Pre- and Post-Combustion Carbon Capture for Power Plants

Ahn, H., University of Edinburgh
Kapetaki, Z., University of Edinburgh
Brandani, P., University of Edinburgh
Luberti, M.,
Brandani, S., University of Edinburgh
Originally presented on: 11/4/2013 12:30:00 - 12:47:00

Process Configuration Studies of Conventional Pre- and Post-Combustion Carbon Capture for Power Plants

Hyungwoong Ahn, Zoe Kapetaki, Pietro Brandani, Mauro Luberti and Stefano Brandani*

Scottish Carbon Capture and Storage Centre (SCCS)

Institute for Materials and Processes, School of Engineering, University of Edinburgh

Mayfield Road, Edinburgh, EH9 3JL, UK

It is well known that an alkanolamine process and a Selexol process are conventional carbon capture processes for post-combustion at the pulverized-coal boiler power plant and for the pre-combustion at IGCC, respectively. The integration of those carbon capture processes into power plants entails considerable energy penalties in power generation.[1] However, the energy penalties resulting from pre- and post- combustion carbon capture can be reduced by a modification of process configuration of amine or Selexol process. A sub-critical PC-fired power plant and an IGCC plant based on a Shell gasifier have been simulated using Honeywell UniSim in order to identify the feed composition to capture units and come up with optimised heat integration of power plant and capture unit. Process simulation of ten different amine processes have been constructed using amine thermodynamic package to estimate the energy saving effect resulting from the improved process configuration[2].

The Selexol process is configured to have two solvent loops to capture H_2S and CO_2 simultaneously. Firstly, it is shown that its energy consumption can be lowered when the two capture loops are integrated rather than separated. It is also proposed that the dual-stage Selexol unit can be further improved by introducing a novel cryogenic capture process for CO_2 bulk separation integrated with a dual-stage Selexol unit for H_2S capture and finishing capture for CO_2 purification[3].

[1] Cost and Performances Baseline for Fossil Energy Plants: Vol.1: Bituminous Coal and Natural Gas to Electricity Final Report, DOE/NETL-2007/1281, 2007.

[2] Ahn, H., Luberti, M., Liu, Z., and Brandani, S., Process Configuration Studies of the Amine Capture Process for Coal-fired Power Plants, Int. J. of Greenhouse Gas Control, Vol. 16, 29-40, 2013.

[3] Kapetaki, Z., Brandani, P., Ahn, H., and Brandani, S., Process Configuration Studies of a Novel Hybrid Cryogenic/Selexol Process Integrated with an IGCC Power Plant, preparing.

(111b) Evaluation of Membrane-Based Post-Combustion Separation of CO2 On Power Plant Output

Swisher, J. A., University of California-Berkeley
Bhown, A. S., Electric Power Research Institute
Originally presented on: 11/4/2013 12:47:00 - 13:04:00

Evaluation of membrane-based post-combustion separation of CO2 on power plant output

Joseph A. Swisher and Abhoyjit S. Bhown

Electric Power Research Institute, Palo Alto, California

Carbon capture, utilization, and storage (CCUS) is an important set of technologies to reduce the carbon dioxide (CO_2) emissions resulting from firing fossil fuels to produce electricity. Due to the large volumes of gas involved, and in the case of post-combustion carbon capture, the low driving force for separation, carbon capture on the scale of a typical fossil fuel-fired power plant is projected to require large capital investment and reduce overall plant efficiency. One benchmark process for CO_2 removal is scrubbing with a solution of monoethanolamine (MEA) which has been estimated to reduce the output of a coal-fired power plant by 30%.[1] Several other gas separation technologies with the promise to reduce the penalty to plant efficiency have been investigated, including advanced solvents, solid sorbents, and CO_2 perm-selective membranes. Membrane-based gas separation processes for post-combustion carbon capture have several potential advantages, including modular scale-up, using no consumable reagents, and not requiring extraction of steam from the plant's power cycle. Proper evaluation of a given membrane technology requires coupling models of membrane performance to models of the rest of the power plant.[2,3,4] In this work, we have developed models of membrane modules from fundamental mass and energy balances. We developed models for all potential flow regimes, co-, counter-, and cross-current, both with and without sweep streams. These models were used to perform parametric studies of separation efficiency under different assumptions about existing and potential membrane properties. These membrane models were linked to a model of a power plant in a process simulator to study the effect of membrane properties and process configurations on the total area required for 90% capture of CO_2 and the attendant effect on net power output.

References

1. Ramezan, M.; Skone, T. J.; Nsakala, N. Y.; Liljedahl, G. N. *Carbon Dioxide Capture from Existing Coal-Fired Power Plants*; DOE/NETL-401/110907; NETL: Pittsburgh, PA, 2007.

2. Merkel, T. C.; Lin, H.; Wei, X.; Baker, R. Power plant post-combustion carbon dioxide capture: an opportunity for membranes. *J. Membr. Sci.* **2010**, *359*, 126-139.

3. Ramasubramanian, K.; Verweij, H.; Winston Ho, W. S. Membrane processes for carbon capture from coal-fired power plant flue gas: A modeling and cost study. *J. Membr. Sci.* **2012**, *421?422*, 299?310.

4. Merkel, T. C. *et al.* Selective exhaust gas recycle with membranes for CO2 capture from natural gas combined cycle power plants. *Ind. Eng. Chem. Res.* **2012**, *52*, 1150-1159.

(111d) Zeolite/Polymer Composite Membranes for CO2 Capture From Flue Gas

Ramasubramanian, K., The Ohio State University
Tong, Z., The Ohio State University
Zhao, L., The Ohio State University
Severance, M. A., The Ohio State University
Dutta, P. K., The Ohio State University
Ho, W. S. W., The Ohio State University
Originally presented on: 11/4/2013 13:21:00 - 13:38:00

Membrane technology provides a promising approach for CO_2 capture from post-combustion flue gas due to its smaller footprint and potentially lower energy cost than the traditional amine scrubbing process. Thin inorganic membranes with high selectivity and flux have been conventionally supported by inorganic materials, which are expensive and difficult for scale-up in a continuous manner and fabrication into a high packing density in terms of high surface area/volume due to their brittle nature.

In this work, a thin zeolite seed layer (< 1 μm) has been successfully deposited on a flexible polymer support for the first time, which enables the continuous roll-to-roll fabrication of the membrane and the configuration of spiral-wound membrane modules with a high packing density. Zeolite-Y with pore size of about 7 Å was specially selected for CO_2/N_2 separation from flue gas, due to the preferential adsorption of CO_2 molecules with a larger quadrupole moment and higher polarizability. This is known as the "surface diffusion" mechanism along with the blocking of N_2 molecules due to the adsorbed CO_2 molecules. In addition, a polymer cover layer containing fixed polyamine and mobile amine carriers was applied on top of the zeolite seed layer by knife-casting to utilize the advantages of both inorganic and polymer membranes for CO_2 capture from flue gas. The thickness of polymer cover layer varied from 0.5 μm to 10 μm. The separation performance of the synthesized zeolite/polymer composite membrane was characterized under humid flue gas composition at close to atmospheric feed pressure and 102°C. The resulting membrane showed a CO_2 permeance greater than 1100 GPU and a CO_2/N_2 selectivity above 400. Based on the transport results, a process and cost modeling study will be discussed in view of achieving the DOE targets of < 35% increase on the cost of electricity and < $40/tonne CO_2 captured for 2025.

Considering the real application in a power plant, SO_2 tolerance of the aforementioned membrane was evaluated under different SO_2 concentrations in feed gas. When SO_2 concentration in feed gas was below 1 ppm, the separation performance of the composite membrane was not affected at all. When 5-ppm SO_2 was supplied, a reduction of CO_2 permeance was observed. However, the CO_2 permeance recovered back to about 80 Â¨C 85% of its original value within 24 hours after exposure to the feed gas without containing the SO_2, which could be crucial if the up-stream SO_2 polishing step fails.

(111e) Membranes for Heat Integrated CO2 Capture Via Sub-Ambient Temperature Operation

Liu, L., Georgia Institute of Technology
Koros, W. J., Georgia Institute of Technology
Sanders, E. S., MEDAL - Air Liquide
Originally presented on: 11/4/2013 13:38:00 - 13:55:00

Pulverized coal (PC) power plants emit CO_2 in flue gas at atmospheric pressure with a volume concentration of 10-15%. Post-combustion capture of CO_2 is a challenging application but has the greatest near-term potential, because it can be retrofitted to existing state-of-the-art power plants. Membrane-based separation is a promising technology for post-combustion CO_2 capture. Our study explored sub-ambient temperature conditions for CO_2/N_2 separation with Matrimid® asymmetric hollow fiber membranes. Due to the lower permeation activation energy for CO_2, a decrease in operating temperature reduced N_2 permeance more than CO_2, resulting in high CO_2/N_2 selectivity. We found that skin morphology had great influence on hollow fiber membrane's separation performance. Fibers with two kinds of skins were successfully spun: fused nodular skins and isotropic dense skins. At -50 °C, the nodular-skinned hollow fiber membranes displayed CO_2/N_2 selectivity higher than 150 with CO_2 permeance higher than 100 GPU. However, the CO_2/N_2 selectivity and CO_2 permeance for dense-skinned samples were much lower. This was hypothesized to be the synergistic effect of increased diffusivity selectivity and solubility selectivity in nodular-skinned hollow fiber membranes at sub-ambient temperatures.

(111f) 3-D Printing As a Tool in the Design of Gas-Liquid Contactors for CO2 Capture

Bara, J. E., University of Alabama
Neuberger, D. T., University of Alabama
Originally presented on: 11/4/2013 13:55:00 - 14:12:00

In order to dramatically reduce CO_2 emissions from coal-fired power plants, and mitigate their impact on global climate change, DOE has called for technologies that can capture at least 90% of CO_2 emissions from an existing coal-fired power plant with no more than a 35% increase in the cost of electricity (COE). One process design approach with the potential to achieve these goals is through the use of advanced gas-liquid contacting devices which enable more efficient capture of $CO_2$with reduced process footprints. However, more cost-effective construction methods and rapid prototyping are needed to deploy these technologies and meet DOE goals.

3-D printing is an "additive" fabrication technique which can offer unprecedented advantages in accelerating the design cycles of gas-liquid contacting devices, minimize manufacturing costs, and expedite deployment timeline for CO_2 capture in the field. Functional prototypes can be designed, fabricated and tested within a frame of 24 hours. Because the design process is entirely software-based, devices can be parametrically varied so that effects of surface area, pressure drop, porosity, etc. can be easily understood and used to develop improved devices. The rapid and flexible feedback loop between design, fabrication and testing that can only be provided through 3-D printing will more quickly advance the peformances and lower the costs of novel gas-liquid contacting devices for CO_2 capture.

This presentation will detail our progress in developing and applying gas-liquid contactors fabricated from 3-D printing techniques. Specifically, we will discuss the advantages, challenges and possibilities relating to solvent-based CO2 capture processes.

(111g) Measurement of Liquid Film Mass Transfer Coefficients for CO2-Binding Organic Liquids (CO2BOLs)

Zheng, R., Pacific Northwest National Lab
Zwoster, A. J., Pacific Northwest National Laboratory
Koech, P. K., Pacific Northwest National Laboratory
Bearden, M., Pacific Northwest National Lab
Freeman, C. J., PNNL
Valkenburg, C., Pacific Northwest National Laboratory
Heldebrant, D. J., Queen's University
Originally presented on: 11/4/2013 14:12:00 - 14:29:00

CO2 absorption kinetics of a CO2BOL solvent were measured using a wetted-wall column. The solvent was a DBU-hexanol two-component first-generation CO2BOL. Liquid film mass transfer coefficients were obtained at 35, 45, and 55°C near atmospheric pressure and three CO2 loading levels at each temperature: 0.5, 1.0, and 4.4 wt% (gram of CO2 per gram of unloaded solvent). The corresponding molar loading levels were 4.5, 9.7, and 43%. On the basis of equal CO2 molar loadings, the data were compared with 5M MEA literature data. Compared to 5M MEA at 40°C and 60°C, at the same molar solvent loading, DBU-hexanol has a larger equilibrium CO_2 pressure P* and lower liquid film mass transfer coefficient (k'_g) values. The general similarity in liquid film mass transfer coefficients (k'g) between MEA and CO2BOLs allows for projections of CO2BOL equipment sizing and costing similar to NETL Case 10 baseline for coal-fired powerplants. Bench-scale testing of CO2BOLs are currently ongoing to provide definitive data on mass transfer for better process economics projections. In this presentation, the above results on the first-generation CO2BOL kinetics will be discussed. Experimental challenges in making accurate WWC measurements on an anhydrous solvent will also be discussed.

(111h) Novel Liquid-Like Organic Hybrid Nanomaterials for CO2 Capture

Park, A. H. A., Columbia University

Originally presented on: 11/4/2013 14:29:00 - 14:46:00

With ever increasing needs for energy independence and sustainability, a myriad of CO_2 capture technologies have been developed including various adsorbents, membranes and solvents. Nanoparticle Organic Hybrid Materials (NOHMs) are one of those novel solvents that have been designed with tunable chemical and physical properties. These nanomaterials are synthesized *via* a straightforward two-step method in which polymeric chains are ionically or covalently tethered onto the surface of inorganic nanocores, leading to the formation of a solvent-free colloidal system. Owing to this unique configuration, NOHMs exhibit enhanced thermal stability and negligible vapor pressure. They can, therefore, address some of the shortcomings encountered by typical amine-based solvents (*i.e.* partial loss of the materials via evaporation and high energy penalty during the solvent regeneration). In this study, several types of NOHMs are synthesized, characterized and evaluated for CO_2 capture under varying conditions. An in-depth investigation of their kinetics, thermodynamics and mechanisms of CO_2 capture in terms of both enthalpic and entropic contributions is conducted. The findings of this research have provided a fundamental knowledge of these novel liquid-like hybrid nanomaterials and have identified key chemical and structural factors ensuring efficient CO_2 capture from conventional power generation and gasification systems.

(111i) The Benefits of Using a Polarity-Swing-Assisted- Regeneration (PSAR) On CO2BOLs: An Experimental and Theoretical Study

Heldebrant, D. J., Queen's University

Originally presented on: 11/4/2013 14:46:00 - 15:03:00

CO_2-Binding Organic Liquids (CO_2BOLs) are a class of second-generation non-aqueous chemically selective CO_2-separating solvents. CO_2BOLs are unique in that in the absence of a co-solvent they are the only solvents that remain a liquid and in the process undergo dramatic changes in polarity with and without CO_2. The solvent's polarity is dictated by the CO_2 loading of the CO_2BOL, and conversely we show here that solvent polarity can control the loading of CO_2. We describe here the application of this unique polarity shift to assist in the release of CO_2 from our CO_2BOL solvents. This polarity-swing-assisted regeneration (PSAR) is conceptually simple, where the addition of a controlled amount of chemically inert non-polar "anti-solvent" is added to the CO_2-rich CO_2BOL with mild heating (>75 °C). Once the CO_2 is released, the antisolvent is separated from the now CO_2-lean CO_2BOL by cooling below the solution's miscibility temperature. The PSAR was found to benefit the CO_2BOL solvents, notably decreasing the reboiler temperatures by as much as 72 °C, and as a consequence, reduce thermal and evaporative losses of the CO_2BOL solvent. The key benefit of a lower reboiler temperature (in anhydrous solvents) is the potential for efficiency gains from the power plant's steam cycle. Preliminary ASPEN plus simulations estimate the addition of the PSAR projects a 37% lower parasitic load compared to DOE's case 10 baseline, with the possibility of 43% when maximizing the polarity assist at full strength. Alternatively, the higher thermal compression of CO_2 can be achieved with the PSAR than conventional thermal regeneration at comparable temperatures; CO_2 pressures as high as 6 ATM have been observed at 100 °C. We present thermodynamic evidence for the PSAR and its effect on VLE and LLE data, followed by ASPEN plus modeling of process configurations and implications. We conclude with projections of net power outputs and parasitic loads of the process as a function of PSAR strength.

(158a) Computational Tools to Accelerate the Development of Carbon Capture Technology

Miller, D. C., U.S. Department Of Energy, National Energy Technology Laboratory

Originally presented on: 11/4/2013 15:15:00 - 15:50:00

The U.S. Department of Energy's Carbon Capture Simulation Initiative (CCSI) was initiated in 2010 to develop and deploy state-of-the-art computational modeling and simulation tools to accelerate the development of carbon capture technology.The first components of the *CCSI Toolset* have been deployed and demonstrated for the design and optimization of post-combustion solid sorbent capture systems. These computational tools provide new capabilities for integrating multi-scale modeling, optimization and uncertainty quantification to rapidly screen new concepts and, ultimately, enable smarter, more focused pilot and demonstration scale projects. The components include a set of detailed process models coupled with a process synthesis and design framework to facilitate development of complete, integrated carbon capture processes using advanced optimization techniques including deterministic superstructure-based methods and derivative-free optimization (DFO). This paper describes these components of the CCSI toolset and how they have been used to develop and optimize a capture process.

(158b) Design of Air Separation Units for Advanced Combustion Via Equation Based Optimization

Dowling, A. W., Carnegie Mellon University
Biegler, L. T., Carnegie Mellon University
Originally presented on: 11/4/2013 15:50:00 - 16:15:00

The quest to develop low carbon emission fossil fuel power systems offers substantial opportunities for systems engineers. Recent advances in optimization algorithms and problem formulations offer a methodology for systematic innovation – through identification of new flowsheets, intensification of existing processes and comparison of promising technologies. Coal oxycombustion is one type of power system with numerous application opportunities for these process systems engineering methods. In this power system coal is combusted in a nitrogen lean environment, resulting in three main products: carbon dioxide, water and pollutants (SO_x, NO_x, PM, Hg, etc). After the pollutants and water are removed, the carbon dioxide is compressed for utilization and/or sequestration. Compared to other low carbon emission coal power systems, oxycombustion shifts the main separation effort to the front of the flowsheet as an $O_2 - N_2$ separation. This avoids the expensive $CO_2 - N_2$separation for large volumes of flue gas.

Coal oxycombustion on a large scale is still a developing technology, and heuristic methods tailored to design of high performance oxycombustion plants do not exist. Furthermore the oxyfuel combustion flowsheet is highly integrated; changing one part of the flowsheet can dramatically impact the performance of other units. This makes design and optimization especially challenging. However this also means the oxycombustion design problem an excellent test-bed to develop new flowsheet optimization technologies.

In this paper we present development of an equation based framework for flowsheet optimization. The framework is designed with highly coupled systems in mind, such as the oxycombustion flowsheet, where optimization with surrogate models may be insufficient. Successful optimization requires coupling detailed process models, including extremely nonlinear thermodynamics, with state-of-the-art optimization algorithms. This framework has several notable features, all tailored to large scale optimization:

- Detailed cubic equation of state thermodynamics models with disappearing (or reappearing) phases
- Approximate distillation models using only continuous variables
- An automated, systematic initialization procedure
- Embedded heat integration based on pinch formulations, evolving with the process optimization
- Exact first and second derivatives

Optimal design of a cryogenic air separation unit for the coal oxycombustion flowsheet is considered to demonstrate the framework. This application is particularly interesting due to the recycle streams and tight heat integration (such as multiple pinch points). Plans to extend the framework to consider other oxycombustion systems will also be discussed.

(158c) Optimal Integration of Compression Heat Within Oxy-Combustion Coal-Based Power Plants

Fu, C., Norwegian University of Science and Technology
Gundersen, T., Norwegian University of Science and Technology
Originally presented on: 11/4/2013 16:15:00 - 16:40:00

Oxy-combustion is a competitive technology to enable the capture of CO_2 from coal based power plants. The reduction in power efficiency and the increment of investment cost are less than (or at least comparable to) other CO_2 capture technologies. The emissions of both SOx and NOx can also be reduced when the oxy-combustion technology is applied [1]. The core concept of oxy-combustion is to use high purity oxygen instead of air for the combustion process so that the flue gas is composed mainly of CO_2 and H_2O. The CO_2 can be separated by condensing the H_2O and then purified by chilling.

The main challenge for implementing the oxy-combustion technology is that a large power penalty is caused by the air separation unit (ASU) and the CO_2 compression and purification unit (CPU). Due to the high volumes of gases to be processed, sub-ambient separation is used in the two units. The thermal efficiency penalty related to CO_2 capture is around 10% points (based on the higher heating value) [2]. Mechanical work is consumed by the compressors for producing the refrigeration energy. In order to reduce the compression work, multi-stage compression with interstage cooling can be applied. An interesting way of utilizing the compression heat is to preheat the boiler feedwater (BFW) in the steam cycle, so that less steam is extracted for the regenerative preheating of the BFW and thus more steam expands through the steam turbines for power generation [3].

When the compression heat is integrated with the steam cycle, multi-stage compression with interstage cooling may not be necessary. The compression heat can be upgraded (the temperature is lifted) by fully or partially adiabatic compression in order to preheat the BFW to a higher temperature. The extraction of steam at the higher pressure levels can be reduced, resulting in more power generated from the steam turbines. On the other hand, however, more work is consumed in the adibatic compression case. Thus there exists a trade-off between the work consumed in the compression processes and the power generated from the steam turbines. Optimal compression schemes should be tailored for the heat integration.

This paper presents a mathematical optimization study of the heat integration. Two challenges will be addressed: (1) how to integrate a heat stream with the steam cycle, and (2) how to optimize the compression schemes. The influences of the compressor efficiency and the temperature differences in the gas/water heat exchangers on the plant performance will be investigated.

References

[1] B.J.P. Buhre, L.K. Elliott, C.D. Sheng, R.P. Gupta, T.F. Wall. Oxy-fuel combustion technology for coal-fired power generation. Progress in Energy and Combustion Science, 31 (2010) 1068-1076.

[2] C. Fu. Process integration in coal based oxy-combustion power plants with CO_2 capture. PhD thesis, Norwegian University of Science and Technology, 2012.

[3] C. Fu, T. Gundersen. Heat integration in coal based oxy-combustion power plants. AIChE Annual Meeting, 28 October - 2 November, 2012, Pittsburgh, USA.

(158d) Performance Analysis of a Staged Pressurized Oxyfuel Combustion (SPOC) Power Plant With Minimal Flue Gas Recycle

Gopan, A., Washington University in St. Louis
Kumfer, B., Washington University in St. Louis
Phillips, J., Electric Power Research Institute
Thimsen, D., Electric Power Research Institute
Axelbaum, R. L., Washington University in St. Louis
Originally presented on: 11/4/2013 16:40:00 - 17:05:00

Power plant CO_2 emissions account for a major portion of the anthropogenic greenhouse gases emitted to the atmosphere. Presently, the costs associated with the capture of carbon dioxide from coal-fired power plants using first-generation oxy-combustion technology are prohibitively high. The U.S. Department of Energy (DOE) has set a goal of developing technologies that can lead to more than 90% capture of carbon dioxide, with an increased cost of electricity of no more than 35%, as compared to a similar plant without carbon capture. One approach that has shown promise to both reduce capital costs and improve power plant efficiency is pressurized oxy-combustion.

Pressurizing the combustion process increases the dew point of the flue gas moisture and allows for condensation at high temperatures. The large latent energy associated with the flue gas moisture can be effectively integrated with the Rankine cycle to increase the efficiency of the process. It also reduces equipment sizes, and eliminates air ingress, resulting in higher CO_2 purity in the raw flue gas.

There have been some studies in recent years to analyze the effect pressurization has on the overall performance of an oxy-combustion power plant [1-3]. Results have shown that an improvement in net efficiency of about 3% points can be gained over what has been reported for first-generation oxy-combustion plants[2, 4].

In this presentation, a new DOE-funded project under the Advanced Oxy-Combustion Technology Development and Scale-up for New and Existing Coal-Fired Power Plants program will be introduced. The goal of the project is to optimize the design of a unique pressurized oxy-combustion process that incorporates fuel staging to minimize the cost of electricity (COE).

A fuel staging approach allows for increased control of the temperature and heat transfer during combustion. This eliminates the need for other temperature control processes, such as flue gas recycle or water/steam injection. The potential benefits of fuel staging which may lead to improved cycle efficiency are: reduced process gas volume, increased average radiation heat transfer, reduced oxygen demands, increased CO_2

purity, and reduced auxiliary power demands.

In the Staged, Pressurized Oxy-Combustion (SPOC) process, combustion is carried out initially at very high stoichiometric ratio, enabling control of temperature even with minimal flue gas recycle. To enable combustion under non-stoichiometric conditions, fuel is brought into the combustion chamber in stages. In the first stage, the coal is burned in nearly pure oxygen. However, the amount of coal is small and the amount of excess O_2 is very large. With proper mixing rates, the temperature of the combustion products of the first stage can be controlled since there is a large amount of excess O_2, which effectively acts as a diluent. Also, the amount of coal consumed in the first stage is relatively small.Â The products of the first stage, a mixture of O_2, CO_2 and H_2O, enter stage 2, where more fuel is injected and more O_2 is consumed. Since the O_2 is diluted with the products and the products have been cooled, the peak flame temperature in this stage can be lower than the first stage. This process continues in multiple stages until nearly all of the O_2 is consumed.

In this work, the performance and costs of a 550MW$_e$ (net) power plant employing the SPOC process are investigated. An integrated pollutant control method was used to remove the NO_x and SO_x. Process and systems modeling was performed using ASPEN Plus software. Sensitivity studies to understand the effect of pressure and the effect of the excess oxygen concentration in the flue gas on the overall performance are also presented.

This study shows that with the SPOC process, a net plant efficiency can be achieved that is similar to that of a sub-critical, air-fired PC power plant that does not have carbon capture. It is also concluded that a more advanced Rankine cycle, with higher throttle pressure and temperature results in an efficiency similar to that of a super-critical, air-fired PC power plant. The increased efficiency, compared to a first-generation oxy-fired plant, is due to both an increase in pressure allowing for the recovery of the latent heat in the flue gas moisture, and the staged process allowing for a significant reduction of the auxiliary load for recycling flue gas.

Acknowledgements

Â· This material is based upon work supported by the Department of Energy under Award Number DE-FE0009702.

Â· Ameren advising team: Rich Smith, Tom Callahan, George Mues.

Â· Engineering, consulting, and cost estimation services provided by Burns & McDonnell.

References:

1. Fassbender, A.G., *Analysis of ThermoEnergy integrated power system for new power facilities using Alaskan and reference coal types*, 2009, AEA.

2. Hong, J., et al., *Analysis of oxy-fuel combustion power cycle utilizing a pressurized coal combustor.* Energy, 2009. **34**(9): p. 1332-1340.

3. Hong, J., *Techno-economic analysis of pressurized oxy-fuel combustion power cycle for CO2 capture*, 2009, MIT: Cambridge.

4. *DOE/NETL-2007/1291. Pulverized Coal Oxycombustion Power Plants. Vol. 1: Bituminous Coal to Electricity, rev. 2*, 2008, U.S. Department of Energy National Energy Technology Laboratory.

(158e) Material Screening and Process Optimization for Cost-Effective Adsorption-Based CO2 Capture

Hasan, M. M. F., Princeton University
First, E. L., Princeton University
Floudas, C. A., Princeton University
Originally presented on: 11/4/2013 17:05:00 - 17:30:00

Elecrticity generation in power plants account for more than 73% of the total stationary CO_2 emissions in the United States [1]. While carbon capture, utilization and storage (CCUS) is an enabling technology to reduce CO_2 emissions from power plants, the costs of CO_2 capture and compression, which represent 60–70% of the total CCUS cost, are estimated to exceed $58/tonne of CO_2 avoided with current technology, leading to about a 63% increase in the levelized electricity cost [2].

Adsorption-based CO_2 capture using zeolites and metal-organic frameworks (MOFs) with large internal surfaces are promising technologies to separate CO_2 from power plant flue gases. In a typical adsorption process, CO_2 is selectively adsorbed onto a solid sorbent while the clean flue gas passes through. The adsorbed CO_2 is released or desorbed by lowering the pressure. While pressure swing adsorption (PSA) process shows promise, the cost of PSA-based CO_2 capture is still high.

We have identified novel materials for cost-effective CO_2 capture by combining *in silico* screening of microporous materials and PSA process optimization for the first time. Our hierarchical computational approach [3-4] efficiently screens large databases of zeolites, based on three dimensional pore characterization [5-6], shape selectivity [7-10], size selectivity, and adsorption selectivity. Next, process optimization is introduced to generate a rank-ordered list based on total cost of capture and compression. The purity, recovery, energy penalty and the cost of capture and compression are obtained by a detailed nonlinear algebraic and partial differential equation (NAPDE) model [11] that describes the PSA process. We not only select the most cost-effective materials, but we also attain the optimal process conditions while satisfying purity, recovery, and other process constraints.

The top zeolites can capture and compress CO_2 to 150 bar from a mixture of 14% CO_2 and 86% N_2 at less than $30 per ton of CO_2 captured and compressed. This is a significant reduction in cost, compared to the costs of absorption and membrane processes [12-14], and other adsorption processes which use zeolite 13X [11]. Several zeolites have moderate selectivities, yet they cost-effectively capture CO_2 with 90% purity and 90% recovery using a 4-step adsorption process. Our results show that no single materials-centric metric is sufficient to always select optimal materials, since both material and process considerations play a role. A combined atomistic, geometric, and process understanding is necessary to achieve the design specifications for a cost-effective and optimal CO_2 capture process.

References:

[1]. DOE/NETL. Carbon Sequestration Atlas of the United States and Canada, 2012.
[2]. Finkenrath, M. I. Cost and performance of carbon dioxide capture from power generation, EIA Report, 2011.
[3]. Hasan, M. M. F.; First, E. L.; Floudas, C. A. Cost-effective CO_2 capture based on in silico screening of zeolites and process optimization. Submitted for publication, 2013.
[4]. U.S. Provisional Patent Application #61/761,436 and #61/765,284.
[5]. First, E. L.; Gounaris, C. E.; Wei, J.; Floudas, C. A. Computational characterization of zeolite porous networks: an automated approach. Phys. Chem. Chem. Phys. 2011, 13:17339-17358.
[6]. First, E. L.; Floudas, C. A. MOFomics: Computational pore characterization of metal-organic frameworks. Micropor. Mesopor. Mater. 2013, 165:32-39.
[7]. Gounaris, C. E.; Floudas, C. A.; Wei, J. Rational design of shape selective separation and catalysis-I: Concepts and analysis. Chem. Eng. Sci. 2006, 61:7933-7948.
[8]. Gounaris, C. E.; Wei, J.; Floudas, C. A. Rational design of shape selective separation and catalysis-II: Mathematical model and computational studies. Chem. Eng. Sci. 2006, 61:7949-7962.
[9]. Gounaris, C. E.; Wei. J.; Floudas, C. A.; Ranjan, R.; Tsapatsis, M. Rational design of shape selective separations and catalysis: Lattice relaxation and effective aperture size. AIChE J. 2009, 56:611-632.
[10]. First, E. L.; Gounaris, C. E.; Floudas, C. A. Predictive framework for shape-selective separations in three-dimensional zeolites and metal-organic frameworks. Langmuir. 2013, 29:5599-5608.
[11]. Hasan, M. M. F.; Baliban, R. C.; Elia, J. A.; Floudas, C. A. modeling, simulation, and optimization of postcombustion CO_2 capture for variable feed concentration and flow rate. 2. Pressure swing adsorption and vacuum swing adsorption processes. Ind. Eng. Chem. Res. 2012b, 51,15665-15682.
[12]. Hasan, M. M. F.; Baliban, R. C.; Elia, J. A.; Floudas, C. A. modeling, simulation, and optimization of postcombustion CO_2 capture for variable feed concentration and flow rate. 1. Chemical absorption and membrane processes. Ind. Eng. Chem. Res. 2012a, 51, 15642-15664.
[13]. DOE/NETL-2010/1397. Cost and performance baseline for fossil energy plants. Vol. 1: Bituminous coal and natural gas to electricity, revision 2. U.S. Department of Energy National Energy Technology Laboratory: Pittsburgh, PA, August 2007.
[14]. Rubin, E.; Zhai, H. The cost of carbon capture and storage for natural gas combined cycle power plants. Environ. Sci. Technol. 2012, 46, 3076-3084.

(158f) Upgrading of Low Rank Coal By The Integration of CCS/Coal-Drying

Park, J. H., Yonsei University
Lee, C. H., Yonsei University
Park, Y. C., Korea Institute of Energy Research
Shun, D., Korea Institute of Energy Research
Bae, D. H., Korea Institute of Energy Research
Park, J., Korea Institute of Energy Research
Originally presented on: 11/4/2013 17:30:00 - 17:55:00

Coal is one of the most important energy sources in the world for electricity and other applications. Coals are generally classified as high and low rank coals depending on their properties, especially heating value, moisture content, impurities etc. Above all, lignite or sub-bituminous coal is a type of low-rank coal (LRC), and it has high moisture content (20% – 50%), lower heating value (3,000 – 4,500kcal/kg on an as-received basis), and high volatile content. These fuels hinder the operation of power plants because the high moisture content reduces the power plant's efficiency and increases flue gas emissions. Also, there are some disadvantages to using LRC such as costly transportation, potential safety hazards during transportation and storage, and operational problems. However, LRC is a competitive primary energy source for power generation. If these fuels are upgraded, those can be used to replace relatively expensive bituminous coals or be blended components with high-rank coal in existing boilers or in new designed boilers. Hence, an appropriate drying process is necessary to develop an energy-efficient power generation systems with lower environmental impacts. Among the other drying systems, the fluidized bed drying system has the advantage of temperature control due to uniformity of bed temperature and high drying rate since the fluidized bed drying involves high heat transfer and mass transfer rates through excellent mixing and gas-solid interactions. It offers a way of drying coal in more economical and environmentally acceptable means.

Recently, many power plants plan to build a CO_2 capture and storage (CCS) process because regulations for greenhouse gas emissions have been reinforced. The CO_2 concentration in earth atmosphere is increased by combusting fossil fuels to generate electricity. Capture of CO_2 in flue-gas streams is essential for the control of CO_2 emissions. Especially, one of the advanced concepts for capturing CO_2 is an absorption process with dry re-generable sorbents. This CO_2 capture process discharges CO_2/steam mixture gas around 150°C of temperature after regeneration reactor. Therefore, we are to develop the fluidized bed dryer using the outlet gas out of the CCS process.

In this study, the purpose of our work is to integrate a CCS and a fluidized bed coal drying system. We produced the upgraded Indonesian coal which has a high moisture content of 35 wt.% using a batch type laboratory-scale fluidized bed dryer.

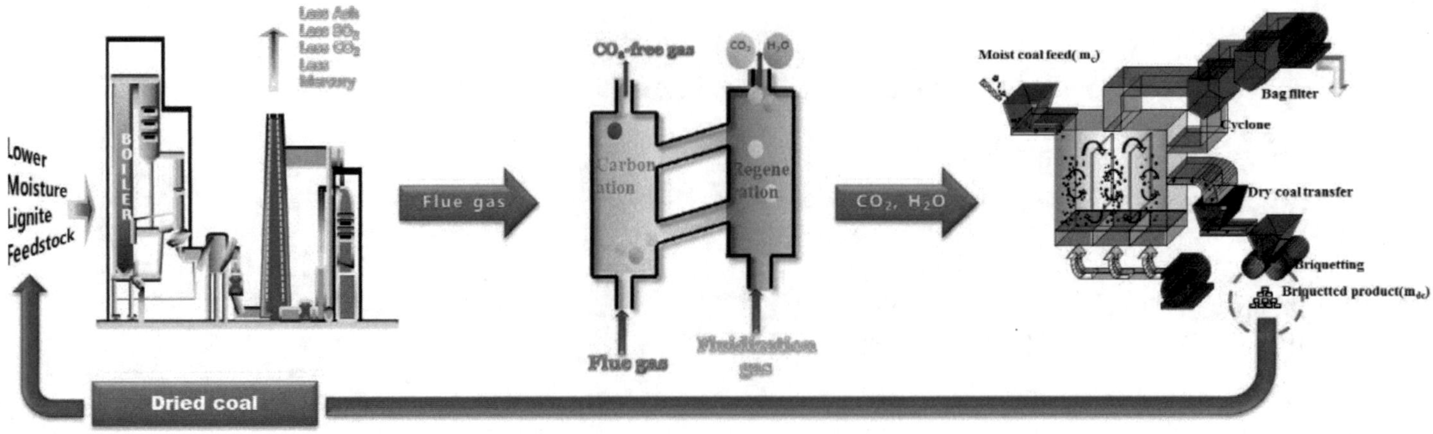

Figure 1. Schematic diagram of integrated CCS/coal-drying system.

Figure 1 shows the concept of integrated CCS/coal-drying system. As shown in the process flow diagram, the CO_2 in the flue gas from fossil fuel-fired power plants is captured through CO_2 capture process. Next, the CO_2 and steam after a CCS process is used as heat sources for drying LRC. Furthermore, CO_2 is used as fluidizing gas and steam is used for indirect coal drying through the heat exchanger. The experiment was performed using CO_2 as a fluidization gas in a fluidized bed dryer (a 500 mm tall pipe with an inner diameter of 80 mm) at a gas velocity ranging from 4 – 6 U_{mf} using a compressed carbon dioxide gas cylinder. The inlet gas temperature was varied from 100 to 200°C using the preheater. The drying rate of the coal was investigated in terms of the inlet gas temperature and the gas velocity in order to determine the optimum operating conditions. Changes in the moisture content of the coal, before and after the experiments, were characterized by a proximate analysis, an ultimate analysis, the higher heating value (HHV), the lower heating value (LHV), a particle size analysis, a thermo gravimetric analysis (TGA). In summary, we are to implement the integrated CCS/coal-drying system using CO_2 and steam as heat sources for LRC drying.

(263a) Hydrogen Production From Natural Gas Via Sorption-Enhanced Variable Volume Batch-Membrane Reactors

Anderson, D. M., Georgia Institute of Technology
Kottke, P. A., Georgia Institute of Technology
Fedorov, A. G., Georgia Institute of Technology
Originally presented on: 11/5/2013 8:30:00 - 8:50:00

Natural gas has become increasingly abundant with recent advances in extraction techniques,[1] motivating the development of new methods for efficient hydrogen production from natural gas in a small-scale, distributed fashion to allow for collocated production/usage and avoid the cost/technical challenges of implementing a hydrogen distribution infrastructure.[2] Steam-methane reforming (SMR) is a mature technology that is the primary route for commercial scale hydrogen production from natural gas.[3] The SMR process consists of three parallel reactions:[4]

(1) $CH_4 + H_2O \leftrightarrow 3H_2 + CO, \Delta H_{298K} = 206.1$ kJ/mol

(2) $CO + H_2O \leftrightarrow H_2 + CO_2, \Delta H_{298K} = -41.2$ kJ/mol

(3) $CH_4 + 2H_2O \leftrightarrow 4H_2 + CO_2, \Delta H_{298K} = 164.9$ kJ/mol

Owing to the strongly endothermic nature of the primary steam reforming reaction, industrial MSR processes typically operate at a highly elevated temperature (800-900°C), followed by low temperature, exothermic water gas shift (WGS) reaction to produce additional H_2.[5] The high temperatures of the industrial MSR reactors are neither practical nor economical for small-scale distributed applications, and the need for subsequent WGS reactors further inhibits straightforward scale-down of the traditional industrial approach. Combined reaction-separation processes, wherein products of the reversible reactions are selectively removed to circumvent equilibrium thermodynamic limitations and enhance forward reaction rates, are an attractive means to reduce the required MSR operating temperature, thus reducing the equipment cost/maintenance expenses and increasing the volumetric yield of hydrogen, both of which are appealing for on-site power generation.[6]

This work explores the potential of the CO_2/H_2 Active Membrane Piston (CHAMP) reactor[7,8] for low-temperature, distributed hydrogen production from natural gas. The CHAMP is a variable volume batch-membrane reactor that operates similar to an Internal Combustion (IC) engine cycle, and is designed to dynamically maintain the optimal conditions (pressure, temperature, and residence time) for transport of reactants to the catalyst, reaction kinetics at the catalyst, and permeation of H_2 through a selective Pd/Ag membrane, as it proceeds through a multi-step reaction-separation cycle. During the reaction step of the CHAMP cycle, judicial compression of the reactor volume is utilized to increase the concentration of the remaining fuel and hydrogen, thereby providing additional driving force for reaction and permeation in order to counteract the reduction in pressure due to permeation and in reaction rate due to fuel depletion.

Herein, thermodynamic analysis is used to show that in order to achieve CH_4 conversion of greater than 90% at temperatures as low as 400°C and at low steam-to-carbon ratios (which are desired for high thermal efficiency and H_2 yield density), a combination of selective CO_2 sorption and H_2 membrane permeation is required. Operational aspects of the sorption-enhanced CHAMP reactor, termed CHAMP-SORB, are discussed, and the thermal efficiency of the cyclic process is reported as a function of the key process variables (pressure, temperature, steam-to-carbon ratio, and sorbent characteristics). The results of the analysis are reported in the form of regime maps for the CHAMP-SORB reactor which can be utilized as design tools to match the reactor's capabilities to desired application requirements. Lastly, the implications of the reaction, permeation and sorption kinetics are also discussed to identify the key timescales governing each step of the CHAMP-SORB cycle and to assess the impact on hydrogen yield/power density of the proposed reactor concept.

References

1. Paltsev S, Jacoby HD, Reilly JM, et al. The future of US natural gas production, use, and trade. *Energy Policy.* 2011;39(9):5309-5321.

2. Barreto L, Makihira A, Riahi K. The hydrogen economy in the 21st century: a sustainable development scenario. *International Journal of Hydrogen Energy.* Mar 2003;28(3):267-284.

3. Baade WF, Parekh UN, Raman VS. Hydrogen. *Kirk-Othmer Encyclopedia of Chemical Technology*: John Wiley & Sons, Inc.; 2000.

4. Xu JG, Froment GF. METHANE STEAM REFORMING, METHANATION AND WATER-GAS SHIFT .1. INTRINSIC KINETICS. *AICHE Journal.* Jan 1989;35(1):88-96.

5. Harrison DP. Sorption-enhanced hydrogen production: A review. *Industrial & Engineering Chemistry Research.* Sep 2008;47(17):6486-6501.

6. Barelli L, Bidini G, Gallorini F, Servili S. Hydrogen production through sorption-enhanced steam methane reforming and membrane technology: A review. *Energy.* Apr 2008;33(4):554-570.

7. Damm DL, Fedorov AG. Batch Reactors for Hydrogen Production: Theoretical Analysis and Experimental Characterization. *Industrial & Engineering Chemistry Research.* Jun 17 2009;48(12):5610-5623.

8. Damm DL, Fedorov AG. Comparative assessment of batch reactors for scalable hydrogen production. *Industrial & Engineering Chemistry Research.* Jul 16 2008;47(14):4665-4674.

(263b) Coupling of Methane Steam Reforming With Chemical Looping Fuel Combustion As a Novel Technology

Rahimpour, M. R., Shiraz University

Originally presented on: 11/5/2013 8:50:00 - 9:10:00

From the environmental perspective, because of the increasing level of CO2 in the atmosphere, different methods are presented for the separating and capturing of CO2 from combustion. One of these new methods is chemical looping combustion (CLC). In this work, the required heat for the endothermic steam reforming process is provided by coupling the CLC as an unprecedented technique with the steam reforming process. Endothermic reactions of methane steam reforming and reduction of Fe-based oxygen carriers in a fuel reactor (FR) are coupled with exothermic reaction of oxygen-carrier oxidation in

an air reactor (AR) of thermally double coupled reactors. In this novel configuration, the AR and FR operate in fast and bubbling fluidization, respectively, and steam reforming occurs in vertically fixed-bed catalytic tubes. The results show that, by application of CLC, not only the amount of produced synthesis gas is the same as the conventional method but also the process whereby CO2 is separated from flue gas inherently. This is the most obvious advantage of the CLC technique. Although the CO2 separation process is very costly, it is necessary for the conventional method. In this study, temperature, mole fraction, methane and air conversion, combustion efficiency, and hydrogen yield have been investigated. Results indicate that the methane conversion and hydrogen yield obtained in this novel method are 29% and 1.04, respectively. Also, the combustion efficiency and gas conversion (O2 and CH4) in both of the CLC reactors (air and fuel reactors) are nearly 100%.

(263c) Bi-Reforming of Methane and Natural Gas From Any Source Under High Pressure. Effect of the Catalyst and Reaction Conditions

Goeppert, A., University of Southern California
Czaun, M., University of Southern California
May, R., University of Southern California
Olah, G., University of Southern California
Prakash, S., University of Southern California
Originally presented on: 11/5/2013 9:10:00 - 9:30:00

Bi-Reforming of Methane and Natural Gas from Any Source Under High Pressure. Effect of the Catalyst and Reaction Conditions

Alain Goeppert, Miklos Czaun, Robert B. May, G. K. Surya Prakash, George A. Olah, Loker Hydrocarbon Research Institute, Department of Chemistry, University of Southern California, Los Angeles, CA 90089, USA

Bi-reforming of methane, the combination of steam and dry reforming, was used to obtain a syn-gas with a H_2/CO ratio of 2 called ?metgas? ideal for methanol synthesis.[1-3] It was achieved by selecting the right proportions between water, CO_2 and CH_4 and reacting them over a suitable catalyst at high temperature.

Steam reforming	$2CH_4 + 2H_2O \longrightarrow 2CO + 6H_2$	$\Delta H_{298K} = 49.1\ kcal.mol^{-1}$	Eq. 1
Dry reforming	$CH_4 + CO_2 \longrightarrow 2CO + 2H_2$	$\Delta H_{298K} = 59.1\ kcal.mol^{-1}$	Eq. 2
Bireforming	$3CH_4 + 2H_2O + CO_2 \longrightarrow 4CO + 8H_2 \longrightarrow 4CH_3OH$		Eq. 3

Bi-reforming to metgas is adaptable for reforming varied natural gas (containing hydrocarbon homologues) and CO_2 sources.

$$3\ C_nH_{(2n-2)} + (3n\text{-}1)H_2O + CO_2 \longrightarrow (3n+1)CO + (6n+2)H_2 \longrightarrow (3n+1)CH_3OH \quad \text{Eq. 4}$$

Bi-reforming could be especially beneficial in the use of natural gas sources containing substantial amounts of CO_2. This CO_2 would, otherwise, have to be separated to allow further processing of the natural gas. Some natural gas sources contain CO_2 concentration from 5% up to 70%. In most cases, once separated, the CO_2 is presently released into the atmosphere.

In order to mimic conditions closer to commercial operations, the bi-reforming reaction was conducted at pressures up to 42 atm in a tubular reactor system specially built for this purpose and able to withstand both high pressure and high temperature in a gas mixture with a high carbon activity. The catalysts used were based on nickel and cobalt deposited on various supports including alumina, alkali earth oxides and combinations thereof. Reaction temperatures ranging from 700Â°C to 910Â°C were investigated. A gas feed composition of $CH_4/CO_2/H_2O$ with a molar ratio of 3/1.2/2.4 was typically used. The reaction was followed by an online GC equipped with a TCD. The amount of carbon formed on the surface of the catalyst was also investigated.

The catalysts were typically tested for at least 100 hours to determine their stability as a function of time. In most cases the conversion of methane as well as carbon dioxide was stable over the length of the experiments (Figure 1). The obtained H_2/CO ratio of the reaction gases was close to the desired value of 2.

The effect of pressure, feed gas composition (amount of water, CO_2, methane), temperature, nature of the catalyst were studied and the results obtained reported.

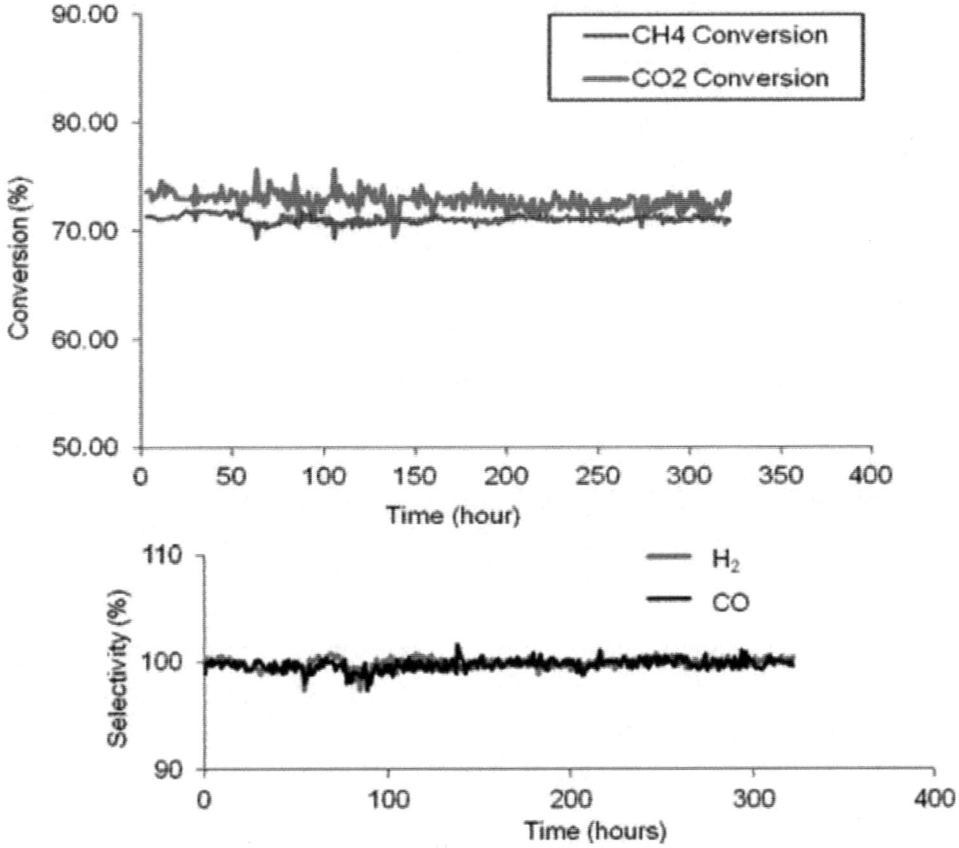

Figure 1. Example of Bi-reforming on 15% NiO on MgO at 7 atm. Conversion of methane and carbon dioxide and selectivity to hydrogen and carbon monoxide.

Acknowledgements

Support of our work by the Loker Hydrocarbon Research Institute and the United States Department of Energy is gratefully acknowledged.

References

[1] G. A. Olah, G. K. S. Prakash, A. Goeppert, *J. Am. Chem. Soc.* **2011**, *133*, 12881.

[2] G. A. Olah, A. Goeppert, G. K. S. Prakash, Beyond Oil and Gas: The Methanol Economy, 2nd ed., Wiley VCH, Weinheim, Germany, **2009**.

[3] G. A. Olah, A. Goeppert, M. Czaun, G. K. S. Prakash, *J. Am. Chem. Soc.* **2013**, *135*, 648

(263d) Activity and Structure of Mo-HZSM5 Catalysts for Non-Oxidative Methane Dehydroaromatization

Natesakhawat, S., National Energy Technology Laboratory
Abdelsayed, V., National Energy Technology Laboratory
Shekhawat, D., National Energy Technology Laboratory
Smith, M., U.S. DOE NETL
Baltrus, J. P., National Energy Technology Laboratory
Sorescu, D. C., National Energy Technology Laboratory
Damodaran, K., University of Pittsburgh
Originally presented on: 11/5/2013 9:30:00 - 9:50:00

Non-oxidative dehydroaromatization (DHA) of methane over Mo/HZSM-5 catalysts offers an attractive route for the production of valuable aromatic compounds such as benzene, toluene, and xylene (BTX). However, there are numerous technical challenges that need to be overcome for developing highly efficient systems. Under typical operating conditions, catalyst deactivation is mainly attributed to coking of the zeolite. Furthermore, there is a long-standing debate in the scientific literature regarding the transformation of MoO_3 to active Mo_xC_y species in these catalysts and deactivation mechanisms. Thus, a better understanding of structure-activity relationships will lead to rational design of DHA catalysts, which in turn will facilitate advances in methane utilization technologies.

We have synthesized and investigated Mo-based catalysts supported on HZSM-5 containing different promoters and varying degrees of acidity. Results from ex situ and in situ surface and bulk characterization techniques have shed light on the factors that control the activity, selectivity, and stability of these catalysts. Correlations between the catalytic performance and Mo_xC_y species associated with the Brönsted acid sites of HZSM-5 will be discussed.

(263e) ZnO@ZSM-5 Nanostructured Core-Shell Catalysts for Methane Dehydroaromatization to Benzene

Lai, Y., University of Pittsburgh
Veser, G., University of Pittsburgh
Originally presented on: 11/5/2013 9:50:00 - 10:10:00

2013 AIChE Annual Meeting
November 3rd - November 8th, 2013
Hilton San Francisco Union Square
California, CA

Abstract Submission

PROGRAM TOPIC:

TE002 Natural Gag rection engineering

Exact Title of Paper:

ZnO@ZSM-5 Nanostructured Core-Shell Catalysts for Methane Dehydroaromatization to Benzene

Presenting Author: Yungchieh Lai

Department of Chemical engineering, University of Pittsburgh, Pittsburgh, PA

Contact Information: yul69@pitt.edu

Co-Authors: Götz Veser

Department of Chemical Engineering, University of Pittsburgh, Pittsburgh, PA

Contact Information: gveser@pitt.edu

Abstract:

Benzene is one of the most important organic intermediates in the US, with an annual volume over 18 billion gallons in 2010 [1]. Currently, most benzene is produced from crude oil. However, increasing crude oil prices and the low cost of natural gas due to the vast amount of accessible shale gas reserves in the US makes natural gas a potential alternative.

The most promising path for methane conversion to benzene is the catalytic dehydroaromatization (DHA) according to: $6CH_4$ à $9H_2 + C_6H_6$

The reaction proceeds over zeolite-based catalysts at ~700-800C with good selectivities (>60%) but low conversion, limited by thermodynamic equilibrium (X_{CH4}~11.8% at 700C, and 21% at 800C). The most widely investigated catalysts for this reaction are bi-functional metal/ZSM-5 catalysts in which the metal site activates methane, followed by oligomerization of the methyl to benzene on the Bronsted Acid Site (BAS) of the zeolite. The pore size of ZSM-5 which is almost identical to the benzene diameter (~ 6 angstrom) results in good selectivity for the desired product benzene.

Among the metals investigated, the most intensively studied one is Mo/ZSM-5 due to its comparatively high reactivity. Membrane reactors used to remove hydrogen in the product was shown to have a further enhancement of methane conversion [2-4]. However, similar to other metal/ZSM-5 systems, this catalyst tends to deactivate due to coke formation especially as produced hydrogen is removed. While this coke formation is reversible (via oxidative regeneration of the catalyst), it results in an oxidized form of the metal which then has to undergo a lengthy activation period to (re)form the active carbide phase. These issues motivate the search for more coking resistant and/or more easily regenerable catalysts for this reaction.

ZnO/ZSM-5 is a possible candidate for this reaction which is active in the oxide form and would hence not require further activation after oxidative regeneration of a spent catalyst. We therefore targeted the synthesis of a nanostructured ZnO@ZSM-5 catalyst with the following desired properties: (1) small ZnO particle size to increase the reactivity of the catalyst, (2) small ZSM-5 particle size and hierarchical pore structure to facilitate removal of the reaction products and hence reduce secondary coke formation.

Here, we report the first successful synthesis of a nanostructured ZnO@ZSM-5 core-shell catalyst system which encapsulates small (~5-10 nm) ZnO nanoparticles in a highly crystalline H-ZSM-5 shell. We will discuss the synthesis path, which is based on a systematic and well-controlled bottom-up approach, and present a detailed characterization of the material. Reactivity and stability tests of the catalyst are planned for the coming future, and will also be included.

References:

[1] B. Balboa. ICIS NEWS 2011.

[2] O. Rival, B. P. A. Grandjean, C. Guy, A. Sayari and F. Larachi. *Ind. Eng. Chem. Res.* 40(2001) 2212-2219.

[3] M. C. Iliuta, F. Larachi, B. P. A. Grandjean, I. Iliuta and A. Sayari. *Ind. Eng. Chem. Res.* 41(2002) 2371-2378.

[4] M. C. Iliuta, B. P. A. Grandjean and F. Larachi. Ind. Eng. Chem. Res. 42(2003) 323-330.

(263f) Spatial Species Concentration and Temperature Profiles in La2O3-CeO2 Nanofiber Catalytic Beds for the Oxidative Coupling of Methane Reaction

Zohour, B., UCLA
Noon, D., UCLA
Senkan, S., UCLA
Originally presented on: 11/5/2013 10:10:00 - 10:30:00

Oxidative coupling of methane (OCM) is a high-temperature process to transform methane into ethylene, which is a valuable intermediate for the chemical industry [1]. In spite of decades long research that resulted in thousands of papers and hundreds of patents, these efforts have failed so far to render OCM as a competitive process. Although many OCM catalysts were reported [2], there appears to be an upper limit of C_{2+} yield of about 25% per reactor pass for which the kinetic reasons are largely unknown [2]. A likely explanation for this limit is that at above a certain temperature, homogeneous gas phase reactions start to dominate over catalysis, so that product selectivities and yields are no longer determined by the catalyst. In order to make progress it is crucial that an improved quantitative understanding of the underlying detailed chemical kinetic mechanisms (DCKM) of the coupled surface and gas phase reactions in OCM be developed *and* validated over a very broad range of conditions encountered in the process. With the availability of DCKM, we will be in a better position to identify improved OCM conditions, superior reactor configurations and new leads for catalytic materials that are needed to exceed the 25% limit for C_{2+} product yields. However, the validation of DCKM require experimental data of high information content since detailed mechanisms contain a large number of species participating in hundreds of elementary reactions. Spatially resolved concentration and temperature profiles over a broad range of conditions represent the necessary and critical data sets for mechanism validation and also represent a new direction in OCM research.

In this presentation, we report on the spatial species concentration and temperature profiles in a fixed bed OCM reactor packed by novel La_2O_3-CeO_2 nanofiber fabric catalysts[3] at different CH_4/O_2 feed ratios at their respected ignition temperatures. The reactor was 6 mm diameter quartz tube and was packed with 20 mg catalyst, which was sandwiched between quartz wool plugs. The concentration profiles were determined by capillary sampling followed by GC analysis. Temperature profiles were determined by thermocouples. The measured mole fraction profiles of CH_4, O_2, C_2H_6, C_2H_6, C_3H_6, CO, CO_2, H_2 and H_2O reveal important quantitative insights on the sequence of reactions of CH_4 and O_2 that are essential for the initial development and validation of detailed chemical kinetic mechanisms for the OCM process.

References

1. M. C. Alvarez-Galvan, N. Mota, M. Ojeda, S. Rojas, R. M. Navarro, J. L. G. Fierro, Catal. Today 2011, 171, 15-23.

2. U. Zavyalova, M. Holena, R. Schlögl, M. Baerns, ChemCatChem 2011, 3, 1935-1947.

3. Noon, D., Seubsai, A., Senkan, S., "Oxidative Coupling of Methane by Nanofiber Catalysis", *ChemCatChem*, **5**, 146-149, 2013.

(263g) Tailored Redox Catalysts for Methane Partial Oxidation

Li, F., North Carolina State University

Originally presented on: 11/5/2013 10:30:00 - 10:50:00

Methane is commercially used as a feedstock for producing hydrogen and liquid transportation fuels via reforming processes. Although a large number of reforming catalysts have been investigated and commercially utilized, deactivation of reforming catalysts remains as a challenge for the development of more effective catalysts. In the current study, we report a "core-shell" redox catalyst that is both an active reforming catalyst and an effective lattice oxygen (O^{2-}) donor for methane reforming. Compared to the conventional reforming processes, the proposed redox catalyst is potentially beneficial since the embedded O^{2-} allows effective oxidation of methane without the presence of external oxidants such as steam or oxygen, avoiding the energy intensive air separation step. The active O^{2-} also inhibits coke formation, allowing a high syngas selectivity without using steam. After O^{2-} donation, the oxygen depleted catalyst can be easily regenerated with steam and/or air. A number of redox catalysts composed of a primary metal oxide and mixed ionic-electronic conductor (MIEC), are synthesized, characterized, and tested under redox conditions. Results indicate that the newly developed core-shell redox catalyst is significantly more selective than conventional oxygen carriers for syngas production. It also exhibits better carbon formation resistance and maintained structural/phase integrity.

(314a) Arsenic Adsorption on Copper-Palladium Alloy Films

Miller, J., USDOE National Energy Technology Department
Uffalussy, K. J., University of South Carolina
Howard, B. H., National Energy Technology Laboratory
Granite, E. J., United States Department of Energy
Stanko, D., U.S. Department of Energy, National Energy Technology Laboratory
Originally presented on: 11/5/2013 12:30:00 - 12:55:00

The adsorption of arsine on copper-palladium alloy films at 288°C was studied by scanning electron microscopy with energy-dispersive X-ray methods (SEM-EDX) characterization. The composition spread alloy film (CSAF) coupons were exposed to arsine in nitrogen in a small flow reactor. The alloys containing predominantly copper (small concentrations of palladium) or predominantly palladium (small concentrations of copper) exhibited large surface concentrations of arsenic after exposure to arsine. The arsenic present on the copper-rich alloy films is consistent with a Cu_3As phase, whereas the arsenic found on the palladium-rich alloy films are consistent with a $PdAs_2$ or $AsPd_2$ phases. These results have potential consequences for the design of sorbents for the capture of arsenic from fluid streams, and for the stability of palladium-copper alloy membranes employed for hydrogen separation from coal-derived syngas.

(314b) On-Line Detection of Elemental and Oxidized Hg Using Quadrupole Mass Spectrometry: Applications and Limits

Rupp, E. C., National Energy Technology Laboratory
Wilcox, J., Stanford University
Originally presented on: 11/5/2013 12:55:00 - 13:20:00

Coal-fired power plants provided between 45 and 50% of the United States net energy generation in the years between 1999 and 2010. While alternate sources of energy, including natural gas and renewable sources, will continue to grow in importance, coal will remain as a necessary item in the energy portfolio of the US. A drawback to coal combustion is the release of mercury, Hg, inherent in the coal matrix. The emissions of Hg was recently regulated by the US EPA in the Mercury and Air Toxics Standards (MATS) ruling in December 2011 and significant efforts are underway to bring power plants into compliance with the regulation. While there are many options for compliance, the standard options, such as activated carbon, rely on the oxidation of Hg and adsorption on a sorbent.

Mercury exists in various forms in the flue gases of coal combustion, from its elemental state (Hg^0) at the boiler exit, to various oxidized forms (Hg^{2+}) as it interacts with fly ash particle surfaces as the flue gas cools. The exact form of the oxidized mercury is, at this point, an outstanding question. Thermodynamic calculations suggest that the favored form in coal combustion flue gas is $HgCl_2$, but HgO and $HgBr_2$ could exist under certain conditions, such as the introduction of Br into the combustion process.

An electron ionization quadrupole mass spectrometer (EI-QMS) has been modified for the direct measurement of elemental and oxidized Hg at industrial relevant concentrations (> 5 $\mu g\ m^{-3}$) from a methane combustion flue gas. Hg^0 and HCl, Cl_2 or Br_2 are introduced pre-combustion and passed through the methane combustion flame to obtain halogen radicals, representative of the real oxidizing environment of a coal-fired boiler. The extent of homogeneous oxidation is measured, and the Hg species are identified through mass spectrometry. In addition, the oxidized species created upon exposure to surfaces present in coal combustion, such as fly ash and activated carbon, are identified through the use of packed bed experiments. A commercial Hg analyzer and speciation system is used to benchmark Hg concentrations determined by mass spectrometry.

(314c) Mercury Oxidation Over Cupric Chloride/Alumina Catalyst for Mercury Emissions Control

Liu, Z., University of Cincinnati
Lee, J. Y., University of Cincinnati
Originally presented on: 11/5/2013 13:20:00 - 13:45:00

Based on our previous mechanistic study of the heterogeneous catalytic oxidation of elemental mercury (Hg(0)) vapor over cupric chloride ($CuCl_2$)-based catalyst on α-Al_2O_3 was conducted in our previous study. The objective of this work is to evaluate the performance of $CuCl_2$/γ-Al_2O_3 under simulated flue gas conditions. Performance tests were conducted in a fixed bed reactor, and both fresh and spent $CuCl_2$/g-Al_2O_3 catalysts were characterized by SEM-EDX, TPD, XRD, XPS, and XAFS.

$CuCl_2$ impregnated on γ-Al_2O_3 showed excellent mercury oxidation performance at 140 °C under a simulated flue gas containing 10 ppbv Hg(0), 10 ppm HCl, 2,000 ppm SO_2, 6% O_2, 7% H_2O, 12% CO_2, and 400 ppm NO. Over 90% Hg(0) oxidation during 150 hours evaluation was achieved and high SO_2 gas concentration (up to 2,000 ppm) did not show any negative effect due to increased acidity on the catalysts surface from $CuCl_2$. Similar to other metal oxide catalysts, NH_3 gas has an inhibition effect on mercury oxidation efficiency. The competitive adsorption of NH_3 gas on active sites may have prohibited the adsorption of reactive species such as HCl. The inhibitory effect of NH_3 gas can be mitigated by increasing HCl concentration. $CuCl_2$ is expected to be able to be used as a catalyst in a temperature window after the air preheater.

(314d) Mercury Oxidation Over Metal Bromide Catalysts for Simultaneous NOx Reduction and Mercury Oxidation

Liu, Z., University of Cincinnati
Lee, J. Y., University of Cincinnati
Originally presented on: 11/5/2013 13:45:00 - 14:10:00

This study aims to develop Hg(0) oxidation catalysts that work in a temperature window (typically 350-400 °C) of the Selective Catalytic Reduction (SCR) unit at a typical coal-fired power plants. Various metal bromides impregnated on TiO_2 and γ-Al_2O_3 were tested in a fixed bed reactor at 350 °C with a simulated flue gas containing 40 ppbv Hg(0), 10 ppm HCl, 2,000 ppm SO_2, 6% O_2, 12% CO_2, 10ppm NH_3 and 400 ppm NO. Metal bromide/TiO_2 catalysts (both fresh and spent) were characterized by SEM-EDX, XAFS, Raman-IR, and XPS. Hg(0) oxidation over metal bromides was found to follow a Mars-Maessen mechanism by consuming lattice bromide which can then be regenerated with the help of HBr and O_2 gases via the Deacon reaction. Metal bromides showed a good tolerance to NH_3 gas for Hg(0) oxidation. Over 95% Hg(0) oxidation was achieved under simulated flue gas conditions, and 10 ppm NH_3 did not show any inhibition effect. In addition, high SO_2 concentration (up to 2,000 ppm) did not show any negative effect on Hg(0) oxidation due to increased acidity on catalyst surface from metal bromides. Metal bromide catalysts can be installed at the tail section of the SCR unit where a NH_3 gas concentration is low. Oxidized mercury in the form of $HgBr_2$ is soluble in water and can be captured in a wet FGD process.

(314e) Modeling of Mercury Capture By CuCl2-Impregnated Activated Carbon Sorbent in Fabric Filter

Li, X., University of Cincinnati
Liu, Z., University of Cincinnati
Lee, J. Y., University of Cincinnati
Originally presented on: 11/5/2013 14:10:00 - 14:35:00

It is estimated that over 300 new fabric filters (FF) are going to be needed for the capture of mercury and fine particulate. We have found that cupric chloride-impregnated activated carbon ($CuCl_2$-AC) demonstrated excellent Hg(0) capture performance. Our previous studies show that mercury adsorption is a sequential process of Hg(0) oxidation and oxidized mercury adsorption over $CuCl_2$-AC. Therefore, it is essential to separately determine the chemical reaction and physical adsorption kinetics for the prediction of Hg(0) vapor removal by $CuCl_2$-AC for sorbent injection.

A model for the physical adsorption of mercuric chloride ($HgCl_2$) onto raw activated carbon (AC), 4% and 10% $CuCl_2$-impregnated activated carbon ($CuCl_2$-AC) sorbents in a fabric filter was studied based on the adsorption equilibrium and kinetics. Sorbent loading, particle size, filtration time, and $CuCl_2$ loading onto AC were found to be the major factors determining an $HgCl_2$ removal efficiency. Although high sorbent loading and small sorbent particle size result in high $HgCl_2$ removal efficiencies, inlet $HgCl_2$ concentration and superficial gas velocity were found to have little impact on $HgCl_2$ removal. Our operation case study result shows that sorbent injection for a short duration at a high loading followed by the discontinuation of the injection until a cleaning cycle can significantly save a total sorbent amount with high sorbent utilization while the same amount of $HgCl_2$ is removed. The adsorption of $HgCl_2$ onto raw AC and $CuCl_2$-AC in the filter cake was found to be primarily governed by the adsorption capacity difference in the Langmuir adsorption kinetic expression. The study results demonstrate how fundamental adsorption equilibrium and kinetics can be used to design a sorbent and predict its performance in a fabric filter.

(314f) Statistical Modelling of Constituents in Ash Evolved From Pulverized Coal Combustion

Jassim, E. I., Prince Mohammad Bin Fahd University
Abdulmouti, H.,
Originally presented on: 11/5/2013 14:35:00 - 15:00:00

Industries using conventional fossil fuels have an interest in better understanding the mechanism of particulate formation during combustion since such is responsible for emission of undesired inorganic elements that directly impact the atmospheric pollution level. Fine and ultrafine particulates have tendency to escape the flue gas cleaning devices to the atmosphere. They also preferentially collect on surfaces in power systems resulting in ascending in corrosion inclination, descending in the heat transfer thermal unit, and severe impact on human health. This adverseness manifests particularly in the regions of world where coal is the dominated source of energy for consumption.

This study highlights the behaviour of calcium transformation as mineral grains verses organically associated inorganic components during pulverized coal combustion. Theinfluance of existing type of calcium on the coarse, fine and ultrafine mode formation mechanisms is also presented. The impact of two subbituminous coals on particle size and calcium composition evolution during combustion is to be assessed. Three mixed blends named Blends 1, 2, and 3 are selected according to the ratio of coal A to coal B by weight. Calcium percentage in original coal increases as going from Blend 1 to 3.

A mathematical model and a new approach of describing constituent distribution are proposed. Analysis of experiments of calcium distribution in ash is also modelled using Poisson distribution. A novel parameter, called elemental index λ, is introduced as a measuring factor of element distribution.

Results show that calcium in ash that originally in coal as mineral grains has index of 17, whereas organically associated calcium transformed to fly ash shown to be best described when elemental index λ is 7.

(360a) Powercells Fuel Processing Development and Its Application in APU System With a PEM Fuel Cell

Ekdunge, P.,
Toftefors, I.,
Kylhammar, L.,
Originally presented on: 11/5/2013 15:15:00 - 15:40:00

Fuel cell systems constitute a clean and efficient alternative to fuel combustion for stationary and mobile applications. One potential market for these systems is Auxiliary Power Units (APUs) for heavy duty vehicles in which an on-board fuel reformer is coupled with fuel cells. Fuel Cell APUs have within the next few years a good chance to reduce fuel consumptions and air pollution as well as improve driver's environment by avoiding the necessity of truck idling. However, despite the potential of these technologies to reduce the environmental impact and improve energy efficiency of road transport, both technical and economic barriers need to be overcome for these technologies to be successfully introduced in mass markets. One challenge is to develop an efficient reformer system that can convert commercial diesel into a hydrogen rich gas stream of sufficient quality for PEM fuel cells.

Development of fuel cell systems and reformer technology for automotive applications started for more than a decade ago within the Volvo Group. PowerCell Sweden AB was spun-out from AB Volvo in 2008 to continue the development and commercialization of fuel cell power systems, diesel-fuel reformers and reformate tolerant LT-PEM fuel cells. During the spring 2013 PowerCell has demonstrated a viable fuel cell APU, based on its proprietary reformer and fuel cell technology. The demonstrated APU is a stand-alone unit that runs on commercial diesel and produces 2.5kW electric energy with 23.5% net efficiency.

The fuel processing for PEM fuel cell APUs requires several steps in order to convert diesel into a hydrogen rich gas stream with low concentration of carbon monoxide (CO) and sulfur species. PowerCell's system consists of an autothermal reformer followed by a desulphurization unit and CO clean-up by using water-gas-shift reaction and preferential oxidation reaction before it can be fed into the fuel cell. The last step in the reformer system is the catalytic burner in which the anode-off gas from the fuel cell is combusted and the released heat is utilized to generate steam for the autothermal reformer.

This paper reviews several reformer generations and solutions developed by PowerCell. The paper also examines what characteristics and requirements of the automotive fuel cell APU that are important to consider in the design and integration of the fuel processor.

(360b) Coupled Operation of a Diesel Steam Reformer and a PEFC

Engelhardt, P., Oel-Waerme-Institut GmbH, Affiliated Institute of the RWTH Aachen University
Maximini, M., Oel-Waerme-Institut GmbH, Affiliated Institute of the RWTH Aachen University
Beckmann, F.,
Brenner, M., Behr GmbH & Co. KG
Originally presented on: 11/5/2013 15:40:00 - 16:00:00

Coupled operation of a diesel steam reformer and a polymer electrolyte fuel cell (PEFC) offers a great potential for auxiliary power units (APU) in mobile applications such as caravans and yachts. Considering its high energy density, diesel is a favourable fuel for early market introduction of mobile fuel cell systems. Furthermore, it is a broadly available fuel in the mobile sector. To generate the hydrogen required for fuel cell operation, on-board fuel processing is necessary.

In a joint research project with partners from industry, Oel-Waerme-Institut GmbH is developing an integrated modular fuel cell system for mobile power generation in caravans and yachts. The system is based on a diesel steam reformer and allows the operation of low-temperature (LT-) as well as high-temperature (HT-) PEFC. In preceding investigations of the author's group, coupling of a steam reformer with an LT-PEFC was demonstrated using a sulfur-free surrogate fuel [1, 2]. Furthermore, the results of a fuel processor optimization regarding start-up, system integration, reformer geometry, and reformer catalyst performance have been reported [2-5].

The objective of this work is to demonstrate the coupled operation of the fuel processor and an LT-PEFC stack using logistic diesel from a gas station as a fuel. The main component of the fuel processor is a catalytically-coated microchannel heat exchanger with a channel height of 0.6 mm provided by BEHR GmbH & Co. KG. The heat exchanger thermally couples the endothermic steam reforming reaction with an exothermic catalytic combustion. The microchannels were alternatingly coated with HyProGen® reformer and burner catalysts, provided by Clariant International Ltd. The reformer catalyst has shown excellent performance with regards to fuel conversion for a sulfur-free diesel surrogate even at reformer temperatures as low as 700 °C. For this work, an LT-PEFC was selected due to its high energy density and excellent start-up performance. A single-stage preferential oxidation reactor was applied to achieve the CO concentration required by the LT-PEFC. Long-term coupled operation was carried out using both sulfur-free diesel surrogate (sulfur content < 2 wt. ppm) and logistic diesel from a German gas station (sulfur content < 10 wt. ppm). Hydrocarbon concentration in the reformate gas at the stack inlet was measured with a flame ionization detector (FID) to investigate the influence on stack voltage for both fuels. During operation with the diesel surrogate, no significant increase in hydrocarbon concentration was observed; the average cell voltage was stable at approximately 630 mV. For logistic diesel operation, an increase in hydrocarbon concentration was measured. Consequently, stack voltage decreased over operating time. With regards to the modular fuel cell system, a strategy for coupled system operation including cyclic catalyst regeneration is presented.

This research and development project is funded by the German Federal Ministry of Economics and Technology (BMWi, FKZ 03ET2052A).

References:

[1] P. Engelhardt, M. Maximini, F. Beckmann, M. Brenner, Int. J. of Hydrogen Energy, 37 (2012) 13470-13477

[2] P. Engelhardt, M. Maximini, F. Beckmann, M. Brenner, World Hydrogen Energy Conference 2012, Toronto, CAN

[3] M. Maximini, P. Engelhardt, M. Brenner, A. De Toni, H.-G. Anfang, World Hydrogen Energy Conference 2012, Toronto, CAN

[4] M. Maximini, P. Engelhardt, M. Grote, M. Brenner, Int. J. of Hydrogen Energy 37 (2012) 10125-10134

[5] M. Grote, M. Maximini, Z. Yang, P. Engelhardt, H. Köhne, K. Lucka, M. Brenner, J. of Power Sources 196 (2011) 9027– 9035

(360c) Evaluation of a Multiple Air Dosage System By Comparing a Monolithic Reactor With a Micro Reactor for Autothermal Steam Reforming of Diesel

Granlund, M. Z.,
GÃ¶rke, O., Karlsruhe Institute of Technology
Pfeifer, P., Karlsruhe Institute of Technology
Pettersson, L. J.,
Originally presented on: 11/5/2013 16:00:00 - 16:20:00

It is estimated that the heavy-duty diesel vehicles (HDDVs) contribute to approximately 30 % of the total emissions of nitrogen oxides (NO_x) and particulate matters (PM) coming from the transport sector [1]. It is estimated that approximately 30 – 40 % of the engine run time is in idling mode [1, 2]. The emission during idling is a large part of the total emission for the vehicle and there is a need to decrease those [3]. One alternative is to use a fuel cell system fueled by diesel, a diesel fuel cell auxiliary power unit (FCAPU), in order to decrease the idling emissions. [4] The FCAPU's main parts are a diesel fuel-processing unit a.k.a reformer, gas clean-up units and a fuel cell.

There are three different processes for reforming fuel into hydrogen-rich gas that can be used; partial oxidation (POX), steam reforming (SR) and autothermal reforming (ATR). The three processes have different advantages and for mobile application ATR is the most beneficial. The main reason is that the process does not need an external heat source, since it is a combination of the exothermic POX and endothermic SR processes. Other advantages are the short start-up time due to the quick thermal response and relatively high hydrogen production. [5]

Due to the limitations in space, onboard HDDV's the FCAPU must have a compact design. The ATR reactor is normally of the tubular type with the catalytic material coated on honeycomb monoliths to minimize the pressure drop, which also makes the reformer resistant to vibrations [6]. One large disadvantage with this reactor design is the characteristic temperature gradient developed over the monolith length, a consequence of the fast POX reaction at the entrance, causing a sharp increase in temperature at the beginning, and the slower steam reforming (SR) towards the outlet of reactor [7]. In order to limit the temperature increase in the beginning, better heat integration between the two reaction zones is necessary. One option to smoothen the temperature gradient could be to use a micro reactor that has better heat integration than the honeycomb monolithic reactor [8]. To further improve the heat integration between the two reaction zones, a micro reactor with four air dosing points axially along the catalytic surface, was designed. The idea is that the heat management could be further improved by dosing the air axially along the micro reactor to adapt the kinetics between the POX and SR reactions via a reactant gradient. With a better heat integration between the POX and SR parts of the catalyst the temperature stress that usually occurs in the beginning of the monolith would be minimized. The integration potentially leads to higher temperatures in the SR part, increasing the rate of reaction and higher H_2yield may be obtained.

The catalytic material used in the comparative study between the two different reactor designs was 3 wt % Rh supported on high surface area CeO_2-ZrO_2 (MEL Chemicals, 16.5 % CeO_2/MOx) prepared by the incipient wetness method ($Rh(NO_3)_3$ solution, 9.53 w/w, Sigma-Aldrich solution). The catalytic material was applied to a 400 cpsi cordierite monolith for the conventional reactor design and on metal foils for the micro reactor design by flow-coating. The fuel used in the comparison between the two reactor designs was biodiesel (NExBTL, Neste Oil, Finland). The main advantages with this biodiesel are its low sulfur and low polyaromatic hydrocarbon content [9].

The monolithic reactor was used to examine the aging pattern and the influence of varying the oxygen-to-carbon and steam-to-carbon ratios in the temperature range of 700 – 900 ¢ªC with external heat source. The experiments were performed at a gas hourly space velocity (GHSV) up to 50 000 $h^{-1}$and the oxygen-to-carbon-ratio was varied between 0.3 and 0.5 and steam-to-carbon ratio varied between 2 and 3.

In the micro reactor the influence of dosing air at the different axial positions was evaluated in the temperature range of 700 – 900 ¢ªC. The amount of air dosed in the four different air inlets were varied in the full interval of 0 – 100 %. By dosing 100 % of the air feed in the first air inlet, the micro reactor's heat integration capacity was evaluated.

Acknowledgement

The financial support from KICInnoEnergy in the project SynCon is gratefully acknowledged. The Swedish Foundation for Strategic Environmental Research (Mistra) is also acknowledged for the financial support. Thanks also to MEL Chemicals, Neste Oil, and Corning Inc. for the supply of the CeO_2-ZrO_2support, NExBTL fuel and cordierite monoliths, respectively.

References

[1] T. Huai, S.D. Shah, J.W. Miller, T. Younglove, D.J. Chernich, A. Ayala, Atmos Environ 40 (2006) 2333-2344.

[2] C.-J. Brodrick, T.E. Lipman, M.Farshchi, H.A.D.N.P. Lutsey, S.W.G.I.D. Sperling, D.B. Harris, F.G.K. Jr., Transp. Res. Part D: Transp. & Environ. 7 (2002) 303-315.

[3] Dieselnet, www.dieselnet.com (accessed 12-10-11).

[4] N. Lutsey, C.J. Brodrick, T. Lipman, Energy 32 (2007-) 2428-2438.

[5] J.D. Holladay, J. Hu, D.L. King, Y. Wang, Catal. Today 139 (2009) 244-260.

[6] C.H. Bartholomew, R.J. Farrauto, Fundamentals of industrial catalytic processes, 2nd ed., Wiley, Hoboken, N.J., 2006.

[7] A. Beretta, A. Donazzi, D. Livio, M. Maestri, G. Groppi, E. Tronconi, P. Forzatti, Catal. Today 171 (2011) 79-83.

[8] J. Thormann, P. Pfeifer, K. Schubert, U. Kunz, Chem. Eng. J. 135 (2008) S74-S81.

[9] Neste Oil, http://www.nesteoil.com (accessed 24-05-13).

(360d) Long-Term Stability At Fuel Processing of Diesel and Kerosene

Pasel, J., Forschungszentrum JÃ¼lich GmbH
Samsun, R. C., Forschungszentrum JÃ¼lich GmbH
Peters, R., Forschungszentrum JÃ¼lich GmbH
Stolten, D., Forschungszentrum JÃ¼lich GmbH
Originally presented on: 11/5/2013 16:20:00 - 16:40:00

Long-term Stability at Fuel Processing of Diesel and Kerosene

The scientific work in the fuel processing and systems group at Forschungszentrum Juelich has the strategic aim of developing a high-temperature polymer electrolyte fuel cell (HT-PEFC) system based on the autothermal reforming of middle distillates such as diesel and kerosene in the power class of 5 to 10 kW_e. This class is particularly interesting for HT-PEFC systems working as auxiliary power units (APUs) for on-board power supply in aircraft and trucks. The main component of the system besides the HT-PEFC stack is the autothermal reformer (ATR). It converts liquid fuels such as diesel or kerosene together with air and steam into a hydrogen-rich gas in a catalytic process (cf. equations (1) and (2)). The exothermic partial oxidation and the endothermic steam reforming reaction of the hydrocarbons contained in diesel fuel and kerosene proceed consecutively on the same monolith. In parallel, the water-gas-shift reaction and the methanation (cf. equations (3) and (4)) reaction also take place in an ATR to the thermodynamic equilibrium of the reactor temperature. Hydrogen, carbon monoxide, carbon dioxide and methane can be regarded as the desired main products of autothermal reforming.

$$C_nH_m + n\,H_2O \quad \text{<---------->} \quad n\,CO + (m/2 + n)\,H_2 \qquad (1)$$

$$C_nH_m + n/2\,O_2 \quad \text{<---------->} \quad n\,CO + m\,H_2 \qquad (2)$$

$$CO + H_2O \quad \text{<---------->} \quad CO_2 + H_2 \qquad (3)$$

$$CO + 3\,H_2 \quad \text{<---------->} \quad CH_4 + H_2O \qquad (4)$$

Reaction equations for autothermal reforming

This presentation will focus on the experimental investigation of the long-term stability of Juelich´s reactor for autothermal reforming ATR 9.2. Experimental results from an experiment for 10,000 h of time on stream using two different synthetic fuels (GTL kerosene and BTL diesel) will be presented, explained and discussed. Special emphasis will be given to the exact determination of hydrocarbon conversion at different points of time during the long-term experiments. For that, the spectra of undesired byproducts and their respective concentrations in the gas and liquid phase were qualitatively determined and quantitatively measured. The spectrum of the byproducts in the gas phase contained short chain hydrocarbons such as ethane, ethane, propene, propane etc. as well as benzene, while that of the liquid phase showed substances such as acetic acid, acetone and butanone. Additionally, the amount of fuel slip if accumulated was gravimetrically measured. After 9,200 h of time on stream for instance, hydrocarbon conversion was still more than 98 %. Furthermore, reaction conditions and operational strategies will be presented and discussed, which allow for at least a partial regeneration of the catalyst.

(360e) Optimization of the Start-Up Strategy for a Diesel Fuel Processor

Maximini, M., Oel-Waerme-Institut GmbH, Affiliated Institute of the RWTH Aachen University
Engelhardt, P., Oel-Waerme-Institut GmbH, Affiliated Institute of the RWTH Aachen University
Brenner, M., Behr GmbH & Co. KG
Anfang, H. G., Clariant Produkte (Deutschland) GmbH
Originally presented on: 11/5/2013 16:40:00 - 17:00:00

Fuel cell systems based on liquid fuels are particularly suitable for auxiliary power generation. For mobile fuel cell systems, high energy density and easy storage of the fuel applied are important. To realize an easy market entry of fuel cell systems, the use of a broadly available fuel like diesel is favorable.

In the joint research project “MÖWE III” Oel-Waerme-Institut and industrial partners are developing a PEFC system based on diesel fuel to be applied as auxiliary power unit (APU) for caravans and yachts. The modular design allows an operation of the fuel processor with LT-PEFC and HT-PEFC providing a maximum power output of 3 kWe. The fuel processor includes a 10 kW microchannel steam reformer provided by Behr GmbH & Co. KG. It thermally couples the endothermic steam reforming reaction with a catalytic combustion using HyProGen® catalysts from Clariant International Ltd. In combination with HT-PEFC, the fuel processor can be coupled directly to the fuel cell. For LT-PEFC operation, a gas cleanup module is integrated.

The objective of the presented work is the optimization of the start-up procedure of the fuel processor. A novel start-up strategy is introduced and demonstrated. The new approach aims at reactive heating of water-gas shift (WGS) reactors and preferential oxidation (PROX) reactors by using reformate from oxidative steam reforming during the reformer heat-up, instead of sequential heating the reactors by large mass flows of hot gases.

The new strategy consists of a sequence of the following phases: preheating, start-up burner operation, reactive reformer heating, oxidative steam reforming, and steam reforming. A complete start-up procedure is demonstrated on a simplified test assembly including two multifunctional reactors for reactant conditioning that are capable to work as: start-up burners, Cool Flame reactors, and diesel-steam mixers. After preheating the 10 kW reformer by combustion gases, oxidative steam reforming is started at approx. 600°C and reformate is obtained from the reformer after 11:15 minutes. Steam reforming is performed after 14:30 minutes at equilibrium concentrations. Until steam reforming operation, 3.07 MJ of Diesel fuel is required. To assess the new procedure in terms of long term operation, soot formation is analysed after 10 of minutes steam reforming operation at 2 kW_{th}.

Eventually, the new procedure is successfully transferred to the complete fuel processor assembly. Concluding, the new start-up procedure is compared to sequential heating of the fuel processor by combustion gases.

Acknowledgement

This research and development project is funded by the German Federal Ministry of Economics and Technology (BMWi, FKZ 03ET2052A)

(360f) Carbon Formation On Pyrochlore Catalysts Under Partial Oxidation of Liquid Hydrocarbons

Smith, M., U.S. DOE NETL
Berry, D. A., National Energy Technology Laboratory
Shekhawat, D., National Energy Technology Laboratory
Haynes, D. J., National Energy Technology Laboratory
Floyd, D.,
Zondlo, J. W., West Virginia University
Spivey, J. J., Louisiana State University
Originally presented on: 11/5/2013 17:00:00 - 17:20:00

Reforming hydrocarbon fuels to produce synthesis gas inevitably involves the formation of carbon. This carbon formation leads to deactivation and is especially problematic for reforming diesel-like fuels under dry oxidation conditions. Rh-substituted pyrochlores (Rh-PC) have high thermal stability and resistance to carbon formation under partial oxidation (POX) conditions with diesel fuel. PC surface properties can be modified through various A- and B-site substitutions; therefore, studying carbon formation on different surface sites could provide insight for catalyst optimization by identifying and promoting the most carbon-resistant surfaces. In this work we report a systematic study of the carbon formation on Rh-PC catalysts with various formulas and surface properties by characterizing the carbon formed in the catalyst bed after varying times on stream in the catalytic POX of a surrogate diesel fuel.

(360g) New Insights On the Role of YSZ in a NiAl2O4/Al2O3-YSZ Diesel Steam Reforming Catalyst

Achouri, E. I.,
Abatzoglou, N., Universite de Sherbrooke
Originally presented on: 11/5/2013 17:20:00 - 17:40:00

A recently patented Ni-Al spinel supported on alumina (Al_2O_3) and Y_2O_3-stabilized ZrO_2 (YSZ) catalyst "$NiAl_2O_4/Al_2O_3$-YSZ" demonstrated high activity and high resistance to coke deposition when used in commercial diesel and biodiesel steam reforming.

In a previous publication it has been shown that the Ni-Al spinel ($NiAl_2O_4$) is deposited preferentially at the alumina phase. Nevertheless, the presence of YSZ had proven instrumental in terms of catalyst efficiency and stability, mainly associated with a decrease of surface carbon formation.In order to explain the YSZ role in the spinel catalyst performance, a wet-impregnation-based protocol was used to prepare catalysts formulations with different Al_2O_3/YSZ mass ratios as well as YSZ with different Y_2O_3/ZrO_2 molar ratios.

Inductively coupled plasma mass spectrometry (ICP-MS) was used to label precisely the prepared catalysts. In all tested formulations the Ni load was kept constant at 5% w/w. The morphology and metal dispersion of the catalysts as well as their crystalline phases, composition and dispersion were examined by scanning electron microscopy coupled with Energy-Dispersive-X-Ray Spectroscopy (SEM-EDXS) and X-ray diffraction (XRD). Temperature-programmed-oxidation and reduction (TPO and TPD) has been used in fresh and used catalytic formulations to evaluate the role of the YSZ as an atomic oxygen-vehicle during the reaction.

The experimental runs are performed in a differential fixed-bed reactor set-up. A proprietary diesel-water emulsion mixture, at low C/H_2O molar ratio of 2 is vaporized in the preheating area of the set-up before entering the steam reforming reaction zone operated at the relatively low -for steam reforming- temperature level of 730°C and mass space velocities ranging between 15 000 and 55 000 $cm^3\ h^{-1}\ g^{-1}$. Gaseous products were analyzed using on-line Gas Chromatography (GC).

The results show that diesel conversion and products selectivity are influenced by the presence of YSZ and its Y (as yttria) content. Moreover, catalyst efficiency over time-on-stream seems being also influenced. More details regarding the mechanism of these phenomena will be presented; they are used to optimize the catalyst formulation.

(360h) Pt/Ni Bimetallic Nanoparticles On Ceria- Based Supports: Hydrogen Production From Autothermal Reforming of Higher Hydrocarbons

Xie, J.,
Walker, B.,
Lawrance, B.,
Sun, X., Kansas State University
Traylor, W., Kansas State University
Hohn, K. L., Kansas State University
Originally presented on: 11/5/2013 17:40:00 - 18:00:00

The catalytic autothermal reforming of hexadecane via Pt/Ni bimetallic nanoparticles on various supports were investigated. Nano particles with Pt/Ni molar ratios ranging from 0 to 10/90 were loaded on ceria-based supports including ceric oxide, gadolium-doped ceric oxide and cerium-doped zirconia oxide. The effect of the Pt/Ni molar ratio and the promotional effect the support were studied by comparing the hydrogen yield and CO, CO_2 and olefin selectivity. TPR and XPS analysis showed the strong interaction between Ni and the Ce-ZrO_x support which lead to the enhancement of catalyst performance when the Pt/Ni ratio is low. With the increase of Pt/Ni mole ratio, bimetallic nanoparticles loaded on Gd-CeO_x exhibit high activity due to the high oxygen mobility of Gd-CeO_x supports.

(424a) Fuel Cell and the Marine & RV Auxiliary Power Market
Finnerty, C.,
Originally presented on: 11/6/2013 8:30:00 - 8:55:00

World-wide development efforts to commercialize portable fuel cells have had limited success. Fuel cell products have struggled with cost-effectiveness, portability, reliability, operational requirements or fuel choice, thus limiting penetration into the portable energy market. WATT Fuel Cell (WATT) has developed what it believes to be the most marketable and cost-effective solid oxide fuel cell (SOFC) technology for portable and potentially small stationary applications. WATT has partnered with Parker Hannifin (Parker) to integrate its next generation fuel cell technology into their combined heat and power (CHP) modular platform. This results in a quiet, reliable and cost effective alternative to generators with the added bonus of heating water which increases system efficiencies further.

WATT's advanced tubular SOFC platform allows for wider operating tolerances which in turn permit the use of inexpensive, yet reliable supporting components to be integrated into the complete system. This also makes vertical integration of WATT's fuel cell system much simpler and effective compared to its competitors. WATT's fuel cell systems are designed based on ease of manufacture, simplicity and cost. These parameters drive decisions on form factor, material choices, reforming chemistry and production processes. WATT's production focused designs provide competitive power densities and overall fuel efficiency for a POX based architecture. A combined heat and power (CHP) system unit using WATT's fuel cell technology will provide customers with new levels of fuel to energy conversion efficiency in the portable energy market creating more value to the customer and a shorter return on investment. WATT's partnership with Parker is expected to bring this fuel efficient and affordable CHP system to the RV and Marine markets and anticipates the market launch to commence in 2014.

WATT's partnership with Parker offers leadership and support during cost reduction strategies and marketing as well as brand-name recognition to new customers. Parker is the world's leading diversified manufacturer of motion and control technologies and systems, providing precision-engineered solutions for a wide variety of commercial, mobile, industrial and aerospace markets. WATT is leading a new era for fuel cell technology focused on cost-optimization of manufacturing fuel cells and related products.

(424b) Hydrogen and Carbon Monoxide Production By Novel Plasma Induced Microreactors

Lindner, P. J., Stevens Institute of Technology
Besser, R. S., Stevens Institute of Technology
Originally presented on: 11/6/2013 8:55:00 - 9:13:00

Hydrogen and Carbon Monoxide Production by Novel Plasma Induced Microreactors

Dr. Peter J. Lindner[1], Dr. Ronald S. Besser[2]

[1]Department of Chemical Engineering

Manhattan College

4513 Manhattan College Parkway

Riverdale, NY 10471

[2]Department of Chemical Engineering and Material Science

Stevens Institute of Technology

One Castle Point on Hudson

Hoboken, NJ 07030

Technological advancements in batteries have not been able to keep pace in the modern era. Despite their low energy densities batteries still remain the only means of portable power. Fuel cells are the logical evolution for a compact and light power source as they have high efficiencies, no moving parts, and can operate at low temperatures. Regardless of these advantages fuel cell technology is not commonly used due to the lack of available hydrogen. Although hydrogen can be produced at large chemical facilities and shipped to local filling stations, the element is very light and takes up a large amount of space even when compressed. The requirement of a large storage volume makes transporting hydrogen impractical for portable applications.

Reforming hydrocarbons locally to feed hydrogen to a fuel cell has been studied to allow for portability. Catalysis of these fuels into hydrogen is commonly used in large scale facilities; scaling them down into a catalytic microreactor seems to be the logical step. Catalysts though suffer from many issues that are exacerbated in the micro-scale. Specifically catalysts require long start up times, high processing temperatures, and become poisoned due to deposits of solid carbon as well as from sulfur components, commonly found with hydrocarbon fuels. Plasma processing has also been studied as a means for locally reforming fuels into hydrogen. Plasma devices typically require high electrical power inputs, making them energy inefficient. Also plasmas generally require additional vacuum equipment, making the devices cumbersome and impractical for portable applications.

A solution that envelops the advantages of both of these systems while avoiding any of their issues is a microplasma reactor. Microplasma reactors are not just miniature plasma reactors, but offer unique advantages such as increased electron density, atmospheric operation, and portability. The microplasma reactors also operate at significantly reduced power requirements. This allows for higher energy efficiency as a more reactive environment is created at a reduced energy cost.

Microplasma reactors have been developed using common microfabricaiton techniques. These devices are produced from a silicon wafer and are smaller than a penny. Previous works from the authors have proven the feasibility of reforming various hydrocarbons into hydrogen such as methane, butane, and methanol but been energy inefficient. Mathematical modeling has shown how simple system modifications could improve efficiencies further. Additionally, an analysis of repeatability of the devices using only carbon dioxide as a feed has demonstrated the decomposition into carbon monoxide, a potential fuel for solid oxide fuel cells.

The microplasma reactors suffer from unknown device failure. This is likely due to electrical shorts, but further rigorous analysis of failed devices is planned. These studies will involve testing the voltage and current characteristics of the terminated device as well as visually inspecting the chips using optical microscopes as well as scanning electron microscopes. The reactors will also be analyzed to determine the chemical composition on the surface of the devices using energy dispersive x-ray spectroscopy.

We plan to present the results of various reforming experiments, our mathematical models, carbon dioxide repeatability tests and device analysis. Future schematics of next generation reactors will also be presented with the aim to develop an all in one fuel processor-fuel cell system.

(424c) Measured and Modeled Properties of Radial Microchannel Reactors (RMR)

Bossard, P., Power and Energy Inc
Breziner, L., Power and Energy
Wilhite, B. A., Texas A&M University
Mettes, J., Power and Energy Inc
Originally presented on: 11/6/2013 9:13:00 - 9:31:00

Efficient catalytic conversion of hydrocarbons to hydrogen and/or synthesis gas remains a critical challenge for both emerging clean energy industries and existing petrochemical and natural gas markets. This paper presents a "Radial Microchannel Reactor" (RMR) architecture developed by Power & Energy, Inc. for manufacturing highly compact steam reformers capable of achieving comparable overall power densities to existing autothermal reformers (ATRs) at reduced catalyst loadings and high space velocities [1]. The RMR system is manufactured via laser welding techniques originally developed by P&E for constructing highly compact gas purification systems. Each microchannel is formed from the annular gap between two nested small tubes, typically designed to have a gap width between 300 and 700 microns. The outer tube is closed at one end and welded into manifold 'A' at its other end. The inner tube has one end, close to the closed end of the outer tube and other end welded to manifold 'B'. More than 1,000 such individual RMR assemblies may be integrated in parallel using a close packed hexagonal pattern in an 8" diameter manifold. This network of RMR assemblies is suspended in a single combustion volume suitable for atmospheric ignition of fuel and air during startup while enabling a smooth transition to catalytic combustion at the outer tube surfaces at steady-state. Catalytic combustion on the outer surface of each RMR provides efficient transfer of combustion heat to the reforming reaction on the inner surface of the outer tube.

This presentation will detail bench-scale experiments for evaluating heat transfer and reaction rates of a single microchannel thermally separated into 3 sections, each 1.25" in length. The inner surface of the larger tube is coated with a nickel-based steam reforming catalyst (~58mg) and an insert tube is coaxially mounted within the bore to complete the RMR geometry. This enables detailed studies of reaction dynamics of the fuel, steam and catalyst issues in a test setup that has the same internal features for the reforming catalyst as a production based RMR system. This apparatus provides heat flow measurements to individual sections of a single microchannel held at constant temperature and is used to measure the power requirements along the outer surface of the individual RMR channel sections to ensure an inner wall temperature of 750°C. Experimental data with methane steam reforming catalyst with a channel gap of 300 microns was obtained over a range of inlet gas space velocities in terms of the net power absorbed by the reforming reaction to maintain reactor-wall temperature. For the single RMR described here, heat duties approaching 40W correspond to a volumetric heat flux of > 800W/cm^3, more than an order of magnitude improvement over competing planar designs. Experimental data is employed to validate computational fluid-dynamic (CFD) models of the RMR system, which are subsequently employed to optimize catalyst distributions and explore thermal integration strategies.

[1] Wilhite, B.A.; L. Breziner; J. Mettes; P. Bossard. Radial Microchannel Reactors (RMR) for Efficient, Compact Steam Reforming of Methane: Experimental Demonstration and Design Simulations, *Energy & Fuels, in press (2013).*

(424d) Development of a Catalytic Micro-Combustor for a Microchannel Methanol Steam Reformer

Montes, M., Faculty of Chemistry
Velasco, I.,
Uriz, I., Universidad Publica de Navarra
Sanz, O., UPV/EHU
Arzamendi, G., Universidad Publica de Navarra
Gandia, L. M., Universidad Publica de Navarra
Odriozola, J. A., University of Sevilla-CSIC
Originally presented on: 11/6/2013 9:31:00 - 9:49:00

Methanol is often considered the best liquid fuel to generate hydrogen due to its high energy density, low cost and easy transportation and reforming. The use of metallic micro-reactors for hydrogen production is gaining an increasing importance regarding portable and low-cost hydrogen sources for numerous applications, particularly for feeding fuel cells. Methanol steam reforming is an endothermic reaction. Therefore, a heat source such as e.g. combustors should be incorporated into the system to maintain reaction temperature and to supply the heat required for the reforming reaction and reactants vaporization. The design, construction and operation of plate-heat-exchanger reactors for application as reformers have been investigated intensively within the last decade [1]. It has been demonstrated that an internal heating by integrated combustion is more efficient than using hot gas from external burners. Methanol combustion can be used because of simplicity.

In the present work, an integrated catalytic microchannel reactor with a micro-combustor coated with Pd/Al2O3 catalyst and a methanol steam micro-reformer coated with Pd/ZnO catalyst was designed and fabricated. The performance of the micro-reformer with the integrated combustor was evaluated under various conditions.

Catalysts were deposited on the microchannel block by washcoating. Several preliminary tests of slurry formulation for washcoating the structural substrates were carried out and the coating characteristics (specific load, homogeneity and adherence) were analyzed to establish the best recipe.

The catalytic tests were carried out at atmospheric pressure in a Computerized Microactivity Reference Catalytic Reactor from PID Eng&Tech. 6 thermocouples were used to monitor temperature: 4 in the same horizontal plane, 2 of them at the microchannels entry and the other 2 at the microchannels outlet, and 2 additional thermocouples at the upper and bottom sides of the block. The shells used to connect the microchannels block also included 4 heating cartridges for the catalyst reduction step. This auxiliary electrical heating system was also used to compensate the heat losses of the microchannel reactor.

The micro-combustor was studied alone using Fecralloy micro-monoliths with differentcombustion catalyst load. In allthe cases total combustion of methanol was achieved, but as methanol/air flow increased, temperature increased as well, and the bulk ofcombustion occurred in the first 10-15 mm ofmonolith.

Based on the results of the combustion monolith, micro-reformer coupled to the combustor was investigated. First, the performance of the micro-reformer without combustion was investigated heating the micro-channel block with the electric cartridges. The methanol reforming conversion was 95 % at 350 ºC and stable selectivity was obtained, producing a verylow CO content (<1 %) in the reformate stream.Measured temperatures were very similar ($\Delta T \leq 5$ ºC)in all points. The power required tomaintain this situation included thepower required to reform the methanol andthe heat lost. In subsequent experiments the power of the cartridges was fixed at decreasing values from initial value to zero and the temperature (350 °C) was controlled through the methanol combustion with a cascade flow- temperature controller. As the electrical power was reduced, the combustion power increased (increased methanol/air flow) producing an increase in the temperature at the combustion channels entry. This increase was due to the fact that the combustion is completed near the entry of the microchannels. Nevertheless, the excellent heat conduction characteristics of the microchannels block homogenized the temperatures.

Acknowledgements

Financial support by the Spanish Ministry of Science and Innovation and Ministry of Economy and Competitiveness (ENE2009-14522-C05 and ENE2012-37431-C03 grants,) and UPV/EHU (GIU11/13) are gratefully acknowledged. I. Uriz and I. Velasco gratefully acknowledge the fellowships granted by the Innovation Department of the Navarre Government and the Spanish Ministry of Science and Innovation (program FPI, BES-2010-030021 and BES-2010-034034).

References

1 G. Kolb, "Fuel Processing for Fuel Cells", Wiley-VCH Verlag GmbH & Co. KGaA Weinheim, Germany (2008).

(424e) Study of the Active Site of Rh Substituted Pyrochlores for Dry Reforming of Methane

Pakhare, D., Louisiana State University
Schwartz, V., Oak Ridge National Laboratory
Haynes, D. J., National Energy Technology Laboratory
Abdelsayed, V., National Energy Technology Laboratory
Shekhawat, D., National Energy Technology Laboratory
Spivey, J. J., Louisiana State University
Originally presented on: 11/6/2013 9:49:00 - 10:07:00

Dry reforming of methane (DRM) to syngas (H_2 and CO) has been well-studied over a number of catalysts. The high temperatures needed to produce syngas inevitably result in carbon formation. Although pyrochlores are known to be thermally stable at conditions required for dry reforming, there is very little literature on this reaction for these catalysts. Catalytically active metals like Rh and Ru can be isomorphically substituted into the crystalline structure of the pryrochlores to produce stable catalyst. These materials also resist carbon formation due to their inherent oxygen conductivity at conditions required for DRM. Here, we report the use of Rh-substituted- lanthanum zirconate ($La_2Zr_2O_7$) pyrochlores for DRM. Specifically three catalysts (a) unsubstituted lanthanum zirconate (LZ) for baseline comparison, (b) 2% and (c) 5% (by wt) Rh-substituted L2RhZ and L5RhZ pyrochlores were characterized using CH_4 and H_2 temperature programmed reduction (TPR) and their activity was studied at different temperatures of 550 °C, 575 °C, and 600 °C. H_2 TPR showed that the reducibility of the pyrochlore increases with an increase in the metal substitution. CH_4 TPR clearly shows that the lattice oxygen in pyrochlores is reactive and helps limit carbon formation. Activity results showed that the CH_4 and CO_2 conversion increase with an increase in active metal substitution into the pyrochlore lattice.

(424f) The Reasoning Behind Long-Term Carbon Free Dry Reforming of Biogas Using Supported Ni-Co Bimetallic Catalysts

Djinovic, P., National Institute of Chemistry
Pintar, A., National Institute of Chemistry
Erjavec, B., National Institute of Chemistry
Osojnik Crnivec, I. G.,
Originally presented on: 11/6/2013 10:07:00 - 10:25:00

Introduction

Biogas, a mixture of methane and carbon dioxide, is formed in large quantities as a result of anaerobic digestion of organic material, wastes and energy crops. Biogas production in Europe has grown exponentially in the last decade, with 10 billion m^3 of biomethane produced in the year 2009, and projected 46 billion m^3 biomethane primary energy potential in 2020 [1]. Composition of biogas is known to vary with time, but is generally rich in methane. It can be combined with CO_2 rich streams, originating for example from thermal power plants to stabilize its chemical composition and thus make it more appropriate for catalytic transformation. Such CH_4/CO_2 gas mixtures can be catalytically upgraded to syngas via the methane dry reforming reaction depicted below.

$$CH_4 + CO_2 \leftrightarrow 2CO + 2H_2 \qquad (1)$$

Syngas produced in this way exhibits the H_2/CO ratio in the ideal case equal to one, which is appropriate for direct synthesis of DME and higher alcohols. These oxygenates are considered as possible fuel substitutes or fuel additives, which can be produced without fossil fuel involvement, thus enabling much cleaner combustion, compared to their non-oxygenated fossil sourced hydrocarbon counterparts. If methane dry reforming reaction is to become a likely pathway for syngas production, catalysts which exhibit good stability and activity are essential. Additionally, they need to be based on transition metals in order to overcome the economic boundaries for their large-scale industrial use. Active catalytic components and their support have to be tailored properly in order to enable stable catalytic performance under the harsh conditions, at which coking of the catalyst surface is thermodynamically favourable.

During this work, we thoroughly investigated how morphology, active metal content and redox properties of various bimetallic NiCo/CeZr catalysts influence their catalytic activity and steered catalyst preparation procedure toward minimal carbon accumulation.

Experimental

CeO_2/ZrO_2 catalyst supports (80:20 by weight) were synthesized by two methods: precipitation with NH_4OH [2] (noted as CeZr HT) and glycothermal method (noted as CeZr EG) [3]. During both protocols, various aging temperatures and times were systematically investigated in order to determine their impact on morphological and chemical properties. Precipitated CeZr nanopowders were calcined 4 h at 400 °C in air. Nickel and cobalt (40:60 by weight) were deposited by homogeneous deposition precipitation method, as suggested by de Jong [4]. The following NiCo loadings were prepared from nitrate precursors: 3, 6, 12 and 18 wt. %. After calcination (4 h at 650 °C), all prepared NiCo/CeZr materials were characterized by N_2 adsorption/desorption, XRD, FE-SEM/EDX and H_2-TPR/TPO methods.

Catalytic activity was evaluated in a quartz fixed-bed reactor (10 mm I.D.), operating at an absolute pressure of 1.2 bar. Prior to these tests, catalyst samples (150 mg, diluted with 850 mg of inert SiC sand) were activated *in-situ* in 20 vol. % H_2/N_2 stream at 750° C for 60 min. CH_4 and CO_2 were fed equimolarly at 50 NmL/min each (GHSV=11000 h^{-1}). The investigated temperature range was between 500 and 800 °C with 50 °C increments.

Results and discussion

Calcined nanocrystalline CeZr HT powders exhibited specific surface areas between 48 and 74 m^2/g, whereas specific surface area of CeZr EG powders up to 121 m^2/g was measured. By increasing aging temperature at fixed aging time, larger CeO_2 crystallites and consequently materials with lower specific surface areas were produced. Prolonging aging time from 6 to 48 h, for the CeZr HT sample with the highest initial specific surface area, revealed a continuous decrease of specific surface area and an increase of average crystallite size from 34.9 to 42.7 nm. On the contrary, average crystallite size of CeZr EG solids was in the range of 5.4 to 14.1 nm.

XRD characterization of prepared solids revealed incorporation of Zr cations into the host CeO_2 lattice, resulting in the formation of homogeneous solid solution; no separated ZrO_2 phase was observed. H_2-TPR characterization of the CeZr HT materials revealed a direct correlation between the CeO_2 crystallite size and higher oxygen storage capacity, which was more pronounced in materials with smaller crystallites. This is probably connected to facile diffusion and a favourable surface-to-volume ratio of nanosized CeO_2 crystallites.

Morphology of NiCo phase was found to be highly dependent on its loading, but not on the synthesis method of CeZr support preparation (HT or EG). At a 3 wt. % loading, nanoparticles were finely dispersed over the CeZr support, while at higher loadings the NiCo phase precipitated preferentially on the NiCo seeds, forming needle-like structures, about 20-30 nm in diameter and 200-500 nm in length. During the catalyst activation step (in which the transformation of nickel on cobalt oxides to NiCo alloy occurred), the NiCo nanowires fused to form round drops from 50 to 500 nm in size. Again, these were larger at higher NiCo loadings. For the 3 wt. % catalyst, however, no change of catalyst morphology after the activation and/or reaction could be identified using SEM microscopy.

Higher dispersion of Ni and Co in catalysts with low metal loadings (3 and 6 wt. %) allowed higher efficiency of active metal particle utilization in the tested reaction. Negligible difference in the morphology was observed between NiCo/CeZr HT and NiCo/CeZr EG catalysts at specific active metal loadings, which translated into very similar catalytic activity. Additionally, this revealed negligible contribution of CeZr morphology directly to methane reforming activity.

Furthermore, significantly different amounts of deposited whisker type carbon on analogous NiCo/CeZr HT and EG catalysts were found to negligibly influence catalytic activity, which was reported previously also by San-JosÃ©-Alonso [5]. The extent of carbon deposition was found to be strongly dependent on NiCo bimetallic particle size and CeZr morphology, which dictates its oxygen mobility.

Oxygen mobility of the support plays a vital role in retarding the carbon accumulation over the tested NiCo/CeZr catalysts. Larger bimetallic particles (NiCo catalysts with higher active metal loadings), enhance coke formation clearly showing that particle size and weaker metal-support interactions enable easier carbon deposition. Secondly, lower oxygen mobility of NiCo/CeZr HT samples enabled less pronounced carbon gasification, which made this regenerative mechanism less efficient. Consequently, the 3-12 wt. % CeZr HT catalysts accumulated an order of magnitude higher carbon quantities compared to analogous CeZr EG catalysts, which exhibited roughly 33 % higher oxygen mobility. The latter could be attributed to the CeO_2 crystallite size, which obviously governs the diffusion of oxygen anions to the surface.

Finally, an optimized NiCo/CeZr EG catalyst (3 wt. % loading) along with the selection of proper operating conditions, enabled to conduct selective reforming (78 and 83 % conversion of CH_4 and CO_2, H_2/CO = 0.85) of undiluted equimolar CH_4/CO_2 stream for more than 400 h, which shows tremendous potential of this material for large-scale applications.

Conclusions

It was discovered that the synthesis method of CeZr support using ethylene glycol with subsequent addition of nickel and cobalt at a loading of 3 wt. % yielded an active methane dry reforming catalyst, which is capable of long-term stable operation in a model biogas stream at 500-750 °C without noticeable coking.

Acknowledgements

The authors gratefully acknowledge the financial support of the Ministry of Education, Science and Sport of the Republic of Slovenia through Research program No. P2-0150.

References

[1] 2011 AEBIOM Annual Statistical Report.

[2] X. Liang, J. Xiao, B. Chen, Y. Li, Inorg. Chem. 49 (2010) 8188.

[3] G.K. Chuah, S. Jaenicke, S.A. Cheong, K.S. Chan, Appl. Catal. A 145 (1996) 267.

[4] K.P. de Jong, Synthesis of solid catalysts, Viley-VCH, Weinheim 2009, pp. 111-132.

[5] D. San-JosÃ©-Alonso, J. Juan-Juan, M.J. Illán-Gómez, M.C. Román-Martínez, Appl. Catal. A 371 (2009) 54.

(424g) In-Situ Probe Technique Applied to Partial Oxidation of Methane and Ethanol On Rh/Al2O3 Coated Monoliths

Diehm, C., Karlsruhe Institute of Technology

Livio, D.,

Hettel, M.,

Torkashvand, B.,

Deutschmann, O., Karlsruhe Institute of Technology (KIT)

Originally presented on: 11/6/2013 10:25:00 - 10:43:00

Catalytic partial oxidation (CPOX) of hydrocarbon fuels to hydrogen consists of a complex coupling of heterogeneous and homogeneous reactions as well as mass and heat transport phenomena. A profound understanding of these processes is crucial for the design and optimization of CPOX reformers. To gain a deeper insight, in-situ sampling techniques are employed [1-3], which yield spatially-resolved concentration and temperature profiles. However, the influence of the probe on the measured data has to be evaluated [4]. Herein, we present experimentally-obtained axial profiles for CPOX of ethanol on a Rh/g-Al_2O_3 coated honeycomb catalyst for various C/O ratios. In addition, different channels of the monolith were examined for CPOX of ethanol and methane. The influence of the probe on the measured data was evaluated by CFD simulations for CPOX of methane.

For the experiments, a monolithic honeycomb (600 channels per square inch (cpsi), Ø 19 mm, L = 10 mm) coated with Rh on a g-Al_2O_3 washcoat was used. Uncoated monoliths (600 cpsi, Ø = 19 mm, L = 10 mm) were placed as heat shields in front of (front heat shield, FHS) and behind the catalyst (back heat shield, BHS), with a gap of 5 mm between the FHS and the catalyst. With a capillary (outer diameter = 170 mm), a constant gas sample was sucked from the channel at the tip of the capillary. The gas composition was analyzed by means of

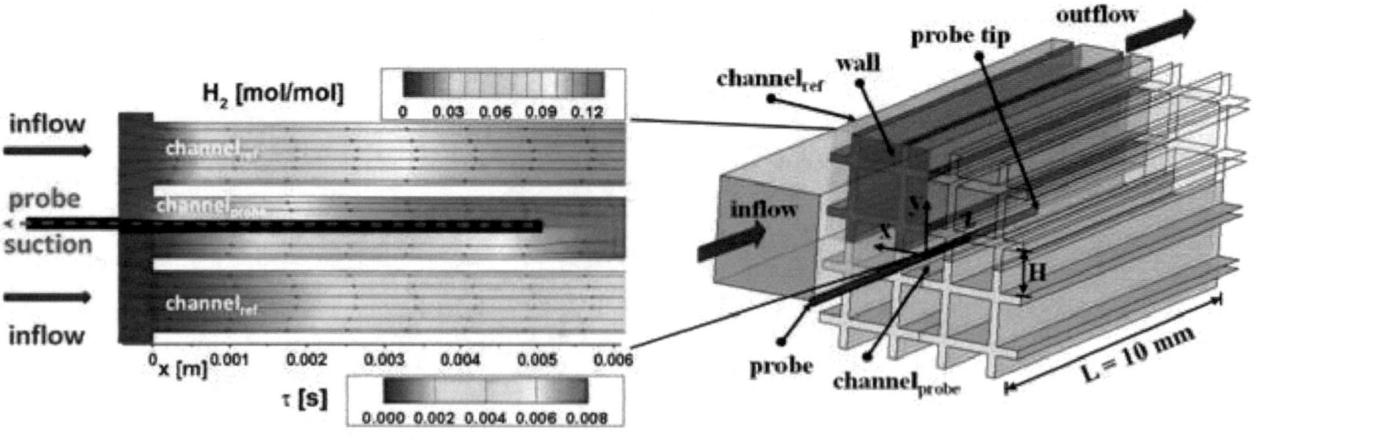

Figure 1: Left: Calculated distributions of H_2 (upper half) and residence time (lower half) in the yz-plane through channel centers for probe tip position z = 5 mm, Right: Calculation domain and boundary conditions.

FT- IR and MS. The temperature profiles were obtained by a thermocouple (gas phase temperature) and an optical fiber connected to a pyrometer (surface temperature).

The computational fluid dynamics (CFD) code FLUENT was applied for the calculations. The computational domain (Figure 1, right) included nine channels of the monolith. Nine different positions of the probe tip inside the catalytic channel were examined. To solve the residence time of the fluid t downstream z = 0, an additional transport equation was implemented into the code. The solid was not included in the simulations; the measured wall temperature was used as a boundary condition.

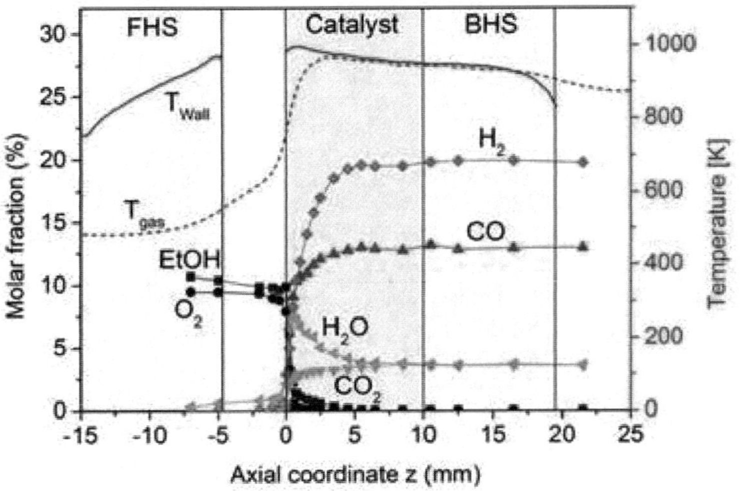

Figure 2: Concentration profiles of reactants and main products (lines with symbols) and temperature profiles of surface (solid line) and gas-phase (dashed line) for CPOX of ethanol (C/O = 0.75).[4]

Typical profiles along a catalytic channel for the concentrations of the reactants and main products are shown for CPOX of ethanol (C/O = 0.75) in Figure 2 (lines with symbols). Two different catalytic zones are observed in the catalytic channel: an oxy-reforming zone with total oxidation and reforming reactions as dominating reactions, and a reforming zone, where all oxygen has been consumed, with only reforming reactions as prevalent reactions. Water (Figure 2) is formed in total oxidation at the catalyst entrance and is partially consumed along the catalytic channel in steam reforming (SR). The surface temperature (Figure 2, solid line) reflects the dominating reactions with a hot spot on the first millimeter inside the channel caused by exothermic total oxidation and a following decrease in temperature due to endothermic reforming reactions, e.g. SR. For ethanol CPOX, pre-catalytic gas-phase reactions, mainly oxidative dehydrogenation of ethanol to water and acetaldehyde (not shown), are observed. The gas-phase temperature (Figure 2, dashed line) increases in front of the catalyst due to heat radiation from the front of the catalyst, which favors the gas-phase reactions.

Figure 1 shows the used computational domain (right) and the calculated distribution of H_2 (left, upper half) and the residence time (left, lower half) for CPOX of methane in a yz-plane starting upstream of the catalyst and ending at 6 mm inside the catalytic channel. Comparing the residence time in a reference channel without

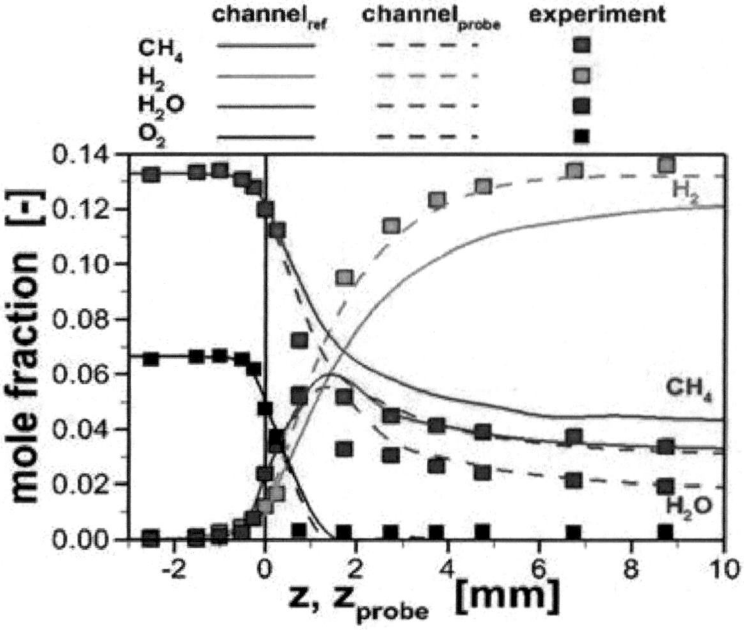

Figure 3: Comparison of calculated (with and without probe) and measured concentrations for the species CH_4, O_2, H_2O, and H_2 as function of axial coordinate for CPOX of methane (C/O = 1.0).[4]

the probe ($channel_{ref}$) with the channel with the probe ($channel_{probe}$), a longer residence time is observed for $channel_{probe}$. Also, a higher hydrogen concentration is observed in the $channel_{probe}$ than in the $channel_{ref}$.

A comparison of the experimentally measured profiles for methane CPOX (symbols) with calculations for a channel with capillary ($channel_{probe}$, dashed lines) and a channel without capillary ($channel_{ref}$, solid lines) is shown in Figure 3.

A deviation between $channel_{ref}$ and $channel_{probe}$ is clearly observable. This is due to the boundary layer, which forms at the outer wall of the capillary. The suction rate does not influence the simulated results, as the sampled gas volume is small compared to the volume flux in the channel. The influence of the probe is dependent on the position of the probe tip in the catalytic channel. The further downstream the probe tip is positioned, the bigger the deviation between $channel_{probe}$ and $channel_{ref}$. The measured data are more similar to the calculated values for a channel with capillary than for one without.

By utilizing the in-situ sampling technique, a deeper insight into the reaction network is gained. The spatially-resolved concentration and temperature profiles reveal the dominant reaction pathways and pre-catalytic reactions like oxidative dehydrogenation in the case of ethanol. However, the influence of the probe on the measured data has to be taken into account. 3D simulations are necessary to evaluate the influence of the probe.

[1] R. Horn, K.A. Williams, N.J. Degenstein and L.D. Schmidt, Journal of Catalysis, 242 (2006) 92.

[2] A. Donazzi, D. Livio, M. Maestri, A. Beretta, G. Groppi, E. Tronconi and P. Forzatti, Angewandte Chemie International Edition, 50 (2011) 3943.

[3] D. Livio, C. Diehm, A. Donazzi, A. Beretta and O. Deutschmann, in preparation.

[4] M. Hettel, C. Diehm, B. Torkashvand and O. Deutschmann, Catalysis Today, submitted.

(424h) Characteristics of Ethylene Generation in the Mixing Region of Autothermal Reforming Applications

Kim, S.,
Bae, J., KAIST
Dean, A. M., Colorado School of Mines
Originally presented on: 11/6/2013 10:43:00 - 11:01:00

Mixture preparation prior to a reforming catalyst is one of the important challenges in developing reliable hydrocarbon reformers. Reactants should be mixed properly before supplying it to the downstream catalyst, and unwanted gas-phase reactions in the mixing region should be suppressed at the same time. Ethylene, a well-known deposit precursor, is one of the major products from gas-phase reactions, which can cause deleterious catalyst failure even with a small amount as ~ 1,000 ppm[1]. Therefore, it is essential to characterize ethylene generation from gas-phase reactions before designing the mixing chamber of a hydrocarbon reformer. Previous works[2-5] showed that gas-phase reactions are closely related to mass and heat transfer; thus, the coupled transport-kinetics was developed to describe the gas-phase reactions in the mixing region of a hydrocarbon reformer[6].

The mixing chamber design described previously[6] was used for the CFD calculations. A mixture of fuel/steam vapor and air are supplied separately and mixed before entering a downstream reforming catalyst. The initial autothermal reforming (ATR) conditions had an O/C ratio of 1.6 and a S/C ratio of 1.5. The flow rates of reactants were calculated with the assumption that the reformer is designed for a $5kW_e$-class system.

N-heptane was selected as a surrogate liquid fuel. The gas-phase kinetic mechanism for *n*-heptane used for the CFD calculations was based on that developed by Westbrook and co-workers[7-9] and consists of 654 species and 2827 reactions. The mechanism was reduced in size because the coupled transport-kinetics model is limited to 50 species. As illustrated in Figure 1, a reduced mechanism consisting of 50 species and 98 reactions was found to be reasonably consistent with the predictions of the original mechanism. This figure shows the decrease in reactivity as the temperature increases from 450 to 550 ^oC. This region of decreasing reactivity with increasing temperature is known as the Negative Temperature Coefficient (NTC) region. The reasons for this unusual behavior are discussed in previous work[10]. Applying the reduced reaction kinetics, the coupled transport-kinetics analysis was performed, and the calculations showed that significant ethylene generation is predicted when the inlet temperature is either below 450 ^oC or above 625 ^oC as described in Figure 2.

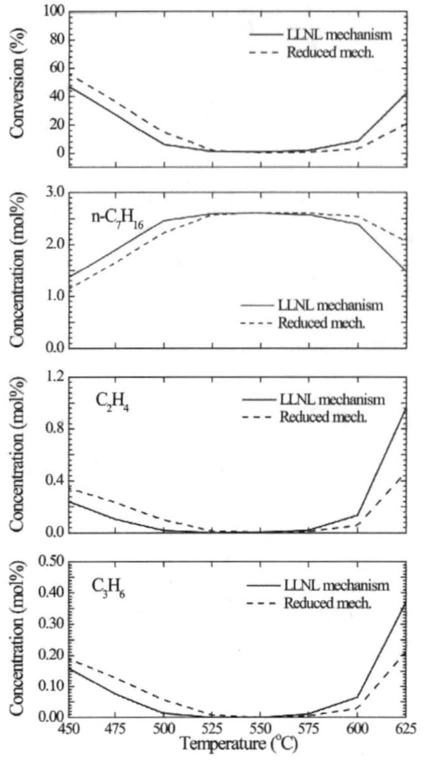

Figure 1 Product distribution of *n*-heptane oxidation modeling results under the nominal ATR condition (O/C=1.6, S/C=1.5): solid line (LLNL mechanism) and dashed line (a reduced mechanism)

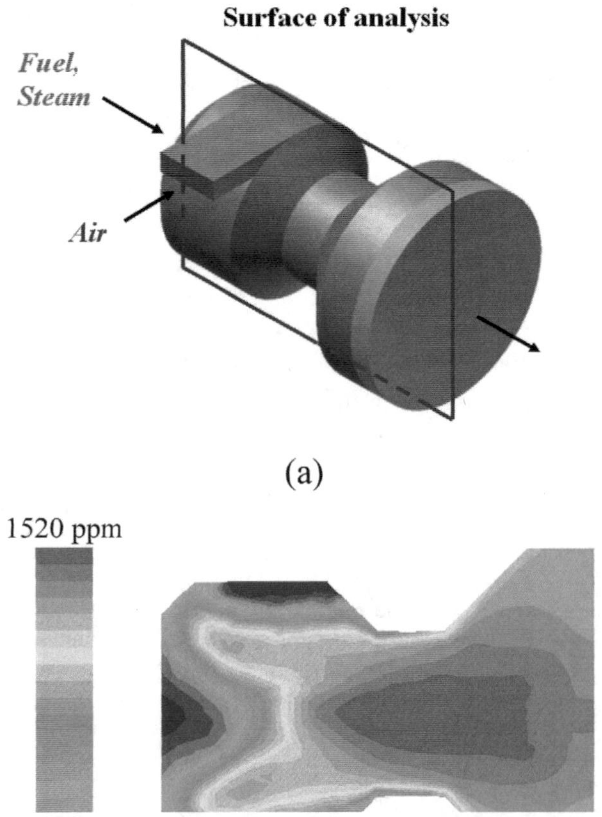

Figure 2 Predicted distributions of ethylene concentration on the cross section shown in (a) at inlet temperature of (b) 450 °C and (c) 625 °C

At lower temperature (450 °C), the local O/C ratio (~1.9) is higher than the overall O/C ratio (1.6) at the location of maximum ethylene production. Therefore, the more oxidizing environment promotes ethylene production. Figure 3 describes the results of the perfectly mixed kinetic study using the reduced mechanism. The rate of ethylene production increases as O/C increases. In this temperature range, most of the ethylene is produced by the reaction $C_2H_5+O_2 = C_2H_4+HO_2$.

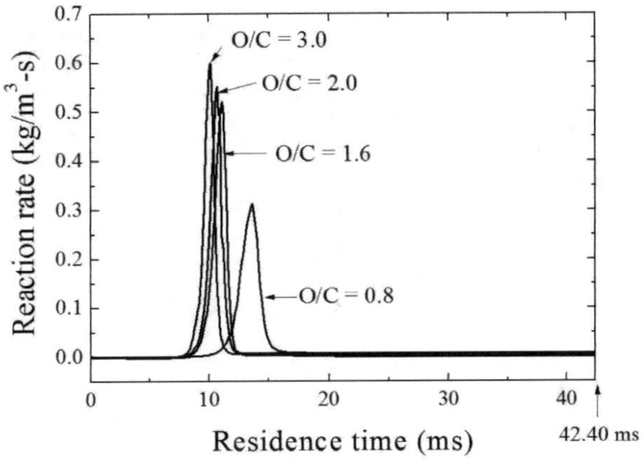

Figure 3 Perfectly-mixed predictions of the reaction rate profiles for ethylene production at the inlet temperature of 450 °C (S/C=1.5)

At higher temperature (625 °C), the rate of ethylene generation and the temperature change are much lower compared to the lower inlet temperature case. The rate of ethylene generation peaks at two points (before and after the nozzle). The results showed that the heat release rate is moderate but coincident with the ethylene generation rate. Figure 4 describes the results of perfectly mixed kinetic calculations also using the reduced mechanism. The effect of O/C is opposite to that predicted at the lower temperature. At the higher temperature, pyrolysis chemistry is dominant, so that the rate of ethylene production decreases as O/C increases. In this temperature range, high-temperature chemistry is activated and alkyl radical decompositions begin to contribute to ethylene production by the following reactions: $n\text{-}C_3H_7 = CH_3+C_2H_4$ and $n\text{-}C_4H_9 = C_2H_5+C_2H_4$. Ethylene production peaks in two locations in the coupled transport-kinetics analysis. At the peak upstream of the nozzle, the local reactant composition is O/C = 0.7. The more pyrolytic environment stimulates the ethylene production; thus the finitely mixed case (266 ppm) showed more ethylene production than the perfectly mixed case (114 ppm).

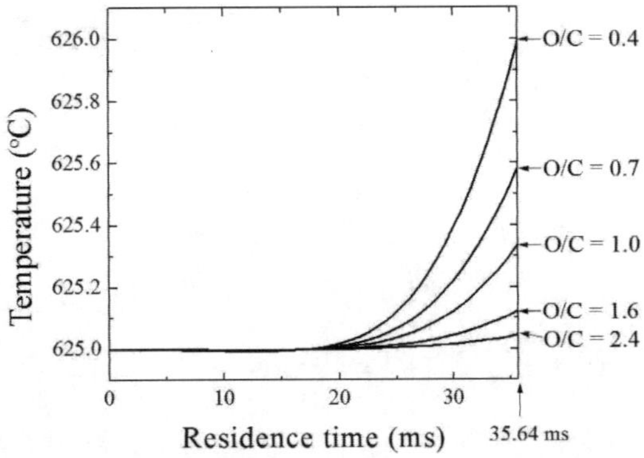

Figure 4 Perfectly-mixed predictions of the reaction rate profiles for ethylene production at the inlet temperature of 625 °C (S/C=1.5)

The coupled transport-kinetics analysis was performed to investigate the characteristics of ethylene production in the mixing region of a hydrocarbon reformer. At lower temperature, most of ethylene is produced by oxygen addition; thus, ethylene production is stimulated with efficient mixing of fuel/steam vapor stream with the air stream. On the other hand, pyrolysis chemistry dominates ethylene production near 625 °C, and ethylene generation can be suppressed by better mixing. The results suggest that the complexities of the gas-phase kinetics demand that careful attention be paid to design of the mixing region upstream of an ATR reformer to avoid undesirable gas-phase reactions that have the potential to interfere with proper catalytic operation.

References

1. Kang IY, Carstensen HH, Dean AM. Impact of gas-phase reactions in the mixing region upstream of a diesel fuel autothermal reformer. Journal of Power Sources. Feb 2011;196(4):2020-2026.

2. Yoon S, Kang I, Bae J. Suppression of ethylene-induced carbon deposition in diesel autothermal reforming. Int. J. Hydrog. Energy. Feb 2009;34(4):1844-1851.

3. Lindstrom B, Karlsson JAJ, Ekdunge P, et al. Diesel fuel reformer for automotive fuel cell applications. Int. J. Hydrog. Energy. May 2009;34(8):3367-3381.

4. Porš Z, Pasel J, Tschauder A, Dahl R, Peters R, Stolten D. Optimised Mixture Formation for Diesel Fuel Processing. Fuel Cells. 2008;8(2):129-137.

5. An SM, Kim WS, Lee SY. Spraying of liquid fuel for improvement of reforming performance for hydrogen generation. Int. J. Hydrog. Energy. 2011;36(9):5342-5349.

6. Kim S, Zhu H, Carstensen HH, Kee R, Dean AM, Bae J. Coupled kinetics and transport in the mixing region of a hydrocarbon reformer. AIChE. Salt Lake City, UT2010:280.

7. Mehl M, Pitz WJ, Sjöberg M, Dec JE. Detailed kinetic modeling of low-temperature heat release for PRF fuels in an HCCI engine. SAE 2009 International Powertrains, Fuels and Lubricants Meeting. Florence, Italy2009.

8. Mehl M, Pitz WJ, Westbrook CK, Curran HJ. Kinetic modeling of gasoline surrogate components and mixtures under engine conditions. Proceedings of the Combustion Institute. 2011;33(1):193-200.

9. Curran HJ, Gaffuri P, Pitz WJ, Westbrook CK. A comprehensive modeling study of n-heptane oxidation. Combustion and Flame. Jul 1998;114(1-2):149-177.

10. Kim S, Carstensen H-H, Dean AM, Bae J. Investigation of gas-phase reactions in the mixing region for hydrocarbon autothermal reforming applications. Int. J. Hydrog. Energy. 2012;37(9):7545-7553.

(487a) Fuel Processing Activities At European Level: A Panoramic Overview

Specchia, S., Politecnico di Torino

Originally presented on: 11/6/2013 12:30:00 - 12:55:00

According to the International Energy Outlook 2011, world marketed energy consumption is growing dramatically, with a worldwide projection rising of 53% from 2008 to 2035, China and India accounting for half of the growth. Expanding energy demand from economic output and improved standards of living will likely put added pressure on energy supplies. On such a context, the European Union's energy approach, the Horizon 2020 strategy, aims to address energy security, resource efficiency and climate change challenges by reducing the greenhouse gas emissions levels by 20%, increasing the share of renewable energies to 20%, and increasing the energy efficiency by 20%

On this point of view, Fuel Cell (FC) technology and fuel processing (FP) of fossil and renewable fuels are playing a crucial role in future sustainable and distributed energy generation for both mobile and portable applications. Specifically, the employment of hydrogen on FCs technology could ensure significant advantages in terms of efficiency and environmental impact, with reduced production of carbon dioxide, representing thus an important alternative to the conventional energy production systems.

Considering the actual lack of infrastructure for hydrogen production, storage and distribution, equipments operating with FCs fed with hydrogen produced by onboard reforming of fossil fuel to generate power or auxiliary power represent a valid and interesting alternative to overcome such an unfavorable situation. In this context, research and development on several reforming systems for fuel processor units (FPUs) has gained a prevalent role in the perspective of solving these problems in a short-medium term. Moreover, FP is a viable option to meet the limited space demands on board vehicles, specifically for auxiliary power units (APUs).

A FP can be realized in a variety of configurations depending on its final application. The first processing step is accomplished by a reforming reactor, which can be a partial oxidation (POX) reactor, or a steam reforming reactor (SR), or an autothermal reformer (ATR). The H_2/CO ratio in the reformate gas depends upon steam-to-carbon (S/C) and oxygen-to-carbon (O/C) ratios, which are a consequence of the fuel to be processed. Subsequent steps are targeted toward reformate conditioning, as adjustment of the H_2/CO ratio and CO removal. When high purity hydrogen is required to feed a low temperature PEM FC, CO has to be almost completely removed thorough a series of catalytic reactors consisting typically of a water gas shift (WGS) reactor and a preferential oxidation (PROX) or selective methanation (SMET) reactor. Usually, an afterburner (AFB) is present, to burn the hydrogen exhaust gas from the PEM-FC anode and provide thus heat to the system. A series of balance of plants (BoP), such as heat exchangers for the internal heat recoveries, water recovery radiators, air compressor, water and fuel pumps, complete the whole FP [1,2].

Considering small and medium size APU systems, ranging from a few kilowatts to a few hundred kilowatts for both portable and mobile applications, microstructured reactors (MSRs) appear to be very promising for achieving maximum compactness. MSRs typically carry small channels, with dimensions in the submillimeter range, with a high surface-area-to-volume ratio, which reduces diffusive transport limitations. Thus, such a technology offers many specific advantages as compared to conventional chemical reactors: (i) enhanced heat transfer, which may be exploited for highly exothermic reactions with the aim of removing the heat generated and suppressing hot spot formation; (ii) superior mass transfer, which can help to optimally control gas concentration profiles in the reactor by suppressing the mass transfer influence on the overall kinetics as well as enhancing the gas adsorption and desorption phenomena at the catalyst surface; (iii) low pressure drop; and (iv) short residence times. When the reactor plates are coated with catalyst, the heat generated by exothermic reactions (or required by endothermic ones) may be removed (or supplied) by design the reactor as a plate heat exchanger, thus improving the thermal management of the reactor itself. Moreover, another way of exploiting the improved heat transfer is the combination of exothermic and endothermic reactions in a single reactor designed like a plate heat exchanger. Thus, the process intensification benefits of microtechnology for gas-phase reactions are currently within the focus of the worldwide research related to fuel processing [3–5].

Many papers available in the literature examined APU systems both from modeling and experimental point of view. As reforming units, SR, ATR, POX, and thermal cracking (TC) have been considered as possible alternative to reformate fossil fuels. On average, SR-based FPs, as compared to the ATR-based ones, showed higher efficiency and larger hydrogen concentration, notwithstanding a slightly higher process scheme complexity and a somewhat more difficult water recovery. As fuels to be reformed, virtually all of the FFs can be considered, starting from gasoline, diesel, biodiesel, propane, and CNG: on average, considering the commercial liquid FFs and by maintaining the same net power released by the APU, the overall efficiency decreased following the rank gasoline, light, and heavy diesel. This is mainly linked to the hydrogen-to-carbon ratio HCR of each fuel: the higher is this ratio (gasoline > light diesel > heavy diesel), the higher is the APU overall efficiency [3,4].

From a technological perspective, actually PEM FCs still need to be considerably improved before penetrating automotive market. Moreover, even considering the doubling of hydrogen refueling stations in Europe in four years, a widespread hydrogen infrastructure does still not exist today. Based on these considerations, reliable, efficient, and quiet APUs are becoming the early market for recreational applications (caravans and yachts), for mobile telecommunication industry (back-up and UPS systems), and for material handling vehicles market (forklifts), with robust commercial products available for business.

The present manuscript provides a panoramic overview of the most recent work carried out at a European level on the research and development of fuel processing technologies for fuel cells. Moreover, an update on actual existing commercial products manufactured in Europe is provided (Helbio, Greece, Truma Gerätetechnik, Germany; UPS, United Kingdom; EPS, Italy; Hygear, The Netherlands).

[1] Cutillo A., Specchia S., Antonini M., Saracco G., Specchia V. J. Power Sources 154 (2006) 379–385.

[2] Kolb G., Cominos V., Hofmann C., Pennemann H., Schurer J., Tiemann D., Wichert M., Zapf R., Hessel V., Lowe H. Chem. Eng. Res. Des. 83 (2005) 626–636.

[3] Specchia S., Specchia V. Ind. Eng. Chem. Res.49 (2010) 6803–6809.

[4] O'Connell M., Kolb G., Schelhaas K.-P., Wichert M., Tiemann D., Pennemann H., Zapf R., Chem. Eng. Res. Des. 90 (2012) 11–18.

[5] Kolb G., Chem. Eng. Proc. Process Intensification 65 (2013) 1–44.

(487b) Effect Of Oxygen Addition To The Feed Of A 5 Kwel Microstructured WATER-GAS Shift Reactors

Schuerer, J., Institut fÃ¼r Mikrotechnik Mainz GmbH
Keller, S., Institut fÃ¼r Mikrotechnik Mainz GmbH (IMM),
Ziogas, A., Institut fÃ¼r Mikrotechnik Mainz GmbH
Thiele, R., Institut fÃ¼r MiKrotechnik Mainz GmbH (IMM)
Kolb, G., Institut fÃ¼r Mikrotechnik Mainz GmbH (IMM)
Originally presented on: 11/6/2013 12:55:00 - 13:13:00

The Water-gas shift reaction is a mandatory reaction to reduce the CO content of reformate produced by decentralized fuel processors dedicated for all kind of PEM fuel cell applications. When adiabatic water-gas shift reactors such as fixed beds or monoliths are applied for this task, this is usually performed in two independent reactor stages with intermediate water-cooling. Plate heat-exchanger technology bears the advantage of combining these two stages to a single reactor (1). When microchannels are applied, the improved heat and mass transfer allows a size reduction of 50% compared to conventional technology (2). The plate heat-exchanger reactor has to be operated with a declining temperature profile, which allows high reaction rate at the reactor inlet and shift of the thermodynamic equilibrium of the reaction towards the outlet (3). Medium temperature water-gas shift catalysts such as platinum/ceria formulations are required to allow the operation at temperatures between 400°C and 250°C, which show high activity and especially high utilization in case microchannels are applied (4). The declining temperature profile is achieved by a coolant gas guided counter-currently through dedicated cooling passages of the reactor. The feasibility of this principle has been demonstrated for microreactors of the kW scale by the group of the authors of the current paper (5, 6). However, the adjustment of the temperature profile is not easily achieved by only cooling the reactor from the product side. The coolant flow rate and temperature mostly determines the reactor exit temperature, while the temperature in the inlet region is dependent on the reformate inlet temperature and the degree of CO conversion in the reactor. The addition of oxygen to noble metal water-gas shift catalysts has been investigated recently as an option to improve catalyst activity at the penalty of certain hydrogen losses owing to unselective combustion (7).

The addition of air to a water-gas shift reactor with an electric power equivalent of 5 kW_{el} was evaluated in the work described by the current paper. It allows the precise adjustment of the reactor inlet temperature and, depending on the catalyst technology applied, increases the CO conversion considerably. On top of that, the air addition is beneficial during transient operation of the reactor and during start-up, the latter because heat can be generated inside the reactor to heat the reactor internally.

(1) G. Kolb, Fuel Processing for Fuel Cells, Wiley-VCH, Weinheim, 2008.

(2) W.E. TeGrotenhuis, D.L. King, K.P. Brooks, B.J. Holladay, R.S. Wegeng, 6th International Conference on Microreaction Technology, IMRET 6, 2002, New Orleans, USA, AIChE Pub. No. 164, 18-28.

(3) T. Baier, G. Kolb, Chem. Eng. Sci. 62 (17) (2007) 4602-4611.

(4) G. Kolb, H. Pennemann, R. Zapf, Catal. Today 110 (1-2) (2005) 121-131.

(5) G. Kolb, C. Hofmann, M. O’Connell, J. Schürer, Catal. Today 147 (2009) 176-184.

(6) M. O'Connell, G. Kolb, K.P. Schelhaas, J. Schuerer, D. Tiemann, A. Ziogas, V. Hessel, Int. J. Hydrogen Energy 35 (2010) 2317-2327.

(7) J. Kugai, J.T. Miller, N. Guo, C. Song, J. Catal. 277 (2011) 46-53.

(487c) Supported Ionic Liquid Phase (SILP) Catalysts for Ambient Pressure and Ultra-Low Temperature Water-Gas-Shift Reaction

Werner, S.,
Wasserscheid, P., Friedrich-Alexander University Erlangen-Nuremberg
Haumann, M., Friedrich-Alexander University Erlangen-Nuremberg
Szesni, N., SÃ¼d-Chemie AG
Kaiser, M., Clariant
Originally presented on: 11/6/2013 13:13:00 - 13:31:00

Supported ionic liquid phase (SILP) catalysts are new materials consisting of an ionic liquid, a metal catalyst and a porous support. The catalyst is dissolved in the ionic liquid which itself is dispersed as a thin film on the inorganic support, thus bridging the gap between traditional homogeneous and heterogeneous catalysis.[1]

Especially continuous, gas-phase reactions are highly suited for this novel and innovative technology. An industrially important example is the water gas shift (WGS) reaction, by which hydrogen can be generated from carbon monoxide and water. Homogeneous WGS catalysts operate at milder temperatures than commercial heterogeneous systems. Since hydrogen production via WGS is an exothermic reaction, lower temperatures result in higher equilibrium conversions.[2]

In this contribution we present investigations of homogeneous metal complexes active in the water gas shift reaction which have been immobilized by the SILP technique. The investigated SILP systems exhibit activities and stabilities exceeding those of homogenous systems reported in literature.[3] Even commercially available WGS catalysts are outperformed, indicating that SILP derived WGS catalysts may become a promising alternative to conventional heterogeneous systems.

[1] a) Riisager, A.; Fehrmann, R.; Flicker, S.; van Hal, R.; Haumann, M.; Wasserscheid, P. *Angew. Chem. Int. Ed.* **2005**, *44*, 185; b) Riisager, A.; Fehrmann, R.; Haumann, M.; Wasserscheid, P. *Eur. J. Inorg. Chem.* **2006**, 695.

[2] Laine, R.M.; Crawford, E.J.; *J. Mol. Cat.* **1988**, *44*, 357.

[3] a) Werner, S.; Szesni, N.; Kaiser, M.; Fischer, R.W.; Haumann, M. Wasserscheid, P. *ChemCatChem.* **2010**, *2*, 1399; b) M. Haumann, A. Schönweiz, H. Breitzke, G. Buntkowsky, S. Werner, N. Szesni, Chem. Eng. Technol. 2012, 35, 1421; c) S. Werner, N. Szesni, M. Kaiser, M. Haumann, P. Wasserscheid, Chem. Eng. Technol. 2012, 35, 1962.

(487d) One-Body Hybrid Solid Packing for Sorption-Enhanced Water Gas Shift Reaction

Lee, C. H., Korea University
Lee, K. B., Korea University
Originally presented on: 11/6/2013 13:31:00 - 13:49:00

The interest in hydrogen, regarded as a new clean energy carrier, has been increasing with the expectation for the upcoming hydrogen economy. In the hydrogen economy, hydrogen can replace the conventional fuels which have caused pollution problems. As one of the methods for mass production of hydrogen, water gas shift (WGS) reaction ($CO + H_2O \leftrightarrow H_2 + CO_2$) has been highlighted for the synthesis gas feed which is produced from coal gasification. Recently, the performance of WGS reaction is significantly increased by applying the sorption-enhanced WGS reaction (SE-WGS) concept, where WGS reaction and CO_2 sorption are carried out simultaneously. High-purity hydrogen can be directly produced through SE-WGS without further purification processes. In the SE-WGS reaction, the uniform packing of catalyst and sorbent is important, however that is difficult to manage with conventional catalyst and sorbent pellets. In this study, novel one-body hybrid solid pellets consisting of the mixture of catalyst and sorbent were prepared to overcome this drawback, and applied to SE-WGS reaction. From experiments, the effect of different ratio of sorbent/catalyst in one-body hybrid solid pellets was studied. A novel multi-section packing concept was also applied to SE-WGS with one-body hybrid solid pellets. The experimental results showed that more high-purity hydrogen could be produced when more catalyst was packed in the first section and more sorbent in the second section.

(487e) Final Step for CO Syngas Clean-Up: Comparison Between CO-PROX and CO-SMET Processes

Ashraf, M. A., Politecnico di Torino
Ercolino, G.,
Vasile, N. S.,
Specchia, S., Politecnico di Torino
Specchia, V., Politecnico di Torino
Originally presented on: 11/6/2013 13:49:00 - 14:07:00

Polymer Electrolyte Fuel Cells (PEM-FC) are the most promising candidate for electric powertrain in transportation applications due to higher energy efficiency and lower greenhouse gas emissions as compared to the internal combustion engines. Hydrogen is normally used as fuel for PEM-FCs. Because of lack of infrastructures and low storage capability on board vehicles, a hydrogen-rich gas can be produced through suitable reforming reactions of fossil fuels followed by water Gas Shift reactors to reduce CO up to 0.5–1%. Considering the actual tolerable level of 10 ppmv CO for PEM-FC's Pt electrocatalyst, the hydrogen-rich gas requires a final CO clean-up step. There are several suitable methods to remove CO from synthesis gas. For PEM-FC application, the low operating pressure processes of preferential CO oxidation (CO-PROX) or CO selective methanation (CO-SMET) are the feasible choice because of the limited spaces available on board vehicles.

CO-PROX reaction has been extensively studied experimentally because of its reliable capability to lower CO to less than 10 ppmv. For example, in a fixed bed reactor the catalyst 1% Rh on zeolite 3A, tested at a WSV of 0.66 NL min^{-1} $g_{cat}{}^{-1}$, was able to reduce the inlet CO concentration below 2 ppmv within a temperature range of at least 80-120 °C [1]. However, this technology requires a closely controlled low O2 supply by expensive mass flow meters, to keep lowest the H2 parallel oxidation, and a wide operating temperature range for control purpose thus it makes costly and complex the process for low power PEM-FC application.

CO-SMET reaction, instead, being able to reduce the CO concentration below 10 ppmv without adding any co-reagent, does not present the above mentioned shortcomings making the process inherently more easily controllable than CO-PROX as, moreover, the selective methanation reaction is less exothermic than CO and H_2 oxidations. For example, in a fixed bed reactor the catalyst 4% Ru γAl_2O_3, tested at a WSV of 0.33 NL min^{-1} $g_{cat}{}^{-1}$, was able to reduce the inlet CO concentration below 2 ppmv within a temperature range of at least 300-340 °C [2]. Even if the H_2 consumption by the selective methanation reaction is higher compared to the hydrogen consumed by the parallel hydrogen oxidation inherently existing in the CO-PROX process, CO-SMET produces methane, inert compound within a fuel cell, but it can be burnt in the afterburner, together with the unreacted anodic hydrogen. In this way, more heat and thermodynamically more useful (since at higher temperature) is generated for the fuel processor compared to heat recovered with the CO-PROX, resulting, therefore, for CO-SMET lower fuel consumption and higher fuel processor efficiency.

Since the two processes for the final CO removal present both pros and cons, the main goal of the present study is to compare and evaluate a Steam Reforming Fuel Processor Unit (SR-FPU) integrated with a PEM-FC for H_2 production for automotive applications, with two different options for the final CO removal: a CO-PROX or a CO-SMET reactor. The integrated SR-FPU consists of a catalytic steam reformer SR, integrated with an afterburner AFB, where the combustion of the H_2 exhaust gas (or H_2 and CH_4 exhaust gas) from the PEM-FC anode takes place to provide the necessary heat to the system. The water gas shift WGS reactor together with CO-PROX catalytic reactor, or CO-SMET catalytic reactor, come to get to completion the transformation of the primary fuel in the secondary one by enabling both the CO clean-up process and the simultaneous increase of the H_2 flow rate. The auxiliary units necessary to properly operate the FPU (heat exchangers for the internal heat recoveries, water recovery radiators, air compressor, water and fuel pumps) complete the BoP of the whole system. Finally, the PEM-FC produces electric power from the secondary fuel. Methane and propane are the considered fuel feedstocks.

The evaluation of SR-FPU performance is done by simulations carried out with a model implemented with the AspenPlus software. The following variables are considered: the steam-to-carbon ratio (SCR = 3.5), a oxygen-to-carbon ratio to avoid coke formation (OCR = 0.07), the SR reactor inlet and outlet temperature (T_{in} = 550 °C, T_{out} = 780 °C), the H_2 utilization coefficient in the PEM-FC anode (η_{AD} = 0.65), the cathode stoichiometry (C_S = 1.8), various efficiencies (FPU and APU), the powers (gross and net) and the overall water balance. Various assumptions were implemented into the model [3,4]: (i) complete conversion of the primary fuel; (ii) the thermodynamic equilibrium in the reformer outlet gas, according to the WGS reaction; (iii) no methanation reaction when the CO-PROX unit is considered; (iv) methanation reaction limited to a value of R ratio (R = mol formed CH_4/mol reacted CO) equal to 1.3 when the CO-SMET unit is considered. The above assumptions are generally confirmed in practice if the reformer volume is sufficiently large and the temperature is adequately high to boost reaction kinetics. As major constrains implemented into the model, the thermal self-sustainability of the SR-FPU and the closure of water balance with no make-up were used. The global thermal power requested to fulfill the T_{in} value of the steam reformer feed is recovered from the hot flow rates of the APU system by suitable heat exchangers. Also the requested water is recovered from the system itself by using suitable air radiators which control the dew point temperature of some gaseous rates rich of water vapor.

Preliminary calculations performed up to now allowed enlightening that the SR-FPU with the CO-SMET reactor presents higher fuel processor efficiency as compared to the SR-FPU with the CO-PROX unit. Details will be provided within the full manuscript.

[1] Galletti C., Specchia S., Saracco G., Specchia V. Ind. Eng. Chem. Res. 47 (2008) 5304-5312.

[2] Galletti C., Specchia S., Saracco G., Specchia V. Chem. Eng. Sci. 65 (2010) 590-596.

[3] Cutillo A., Specchia S., Antonini M., Saracco G., Specchia V. J. Power Sources 154 (2006) 379-385.

[4] Specchia S.; Cutillo A., Saracco G., Specchia V. Ind. Eng. Chem. Res.45 (2006) 5298-5307.

(487f) Planar and Modular Microstructured Reactor With Integrated Palladium Membrane for On-Site Production of Pure Hydrogen

Boeltken, T., Karlsruhe Institute of Technology
Lee, S., Karlsruhe Institute of Technology
Pfeifer, P., Karlsruhe Institute of Technology
Peters, T., SINTEF
Bredesen, R., SINTEF
Dittmeyer, R., Karlsruhe Institute of Technology
Originally presented on: 11/6/2013 14:07:00 - 14:25:00

Hydrogen is one of the most important chemicals used in industry today. Over 80% of today's hydrogen demand is covered by steam reforming of natural gas in large units with up to 1000 tubular reformers and capacities up to 100.000 Nm^3/h [1]. Furthermore, hydrogen is often considered as the energy carrier of the future and its application as fuel for fuel cell vehicles and distributed power systems, which could reduce local CO, NO_x and CO_2 emissions, is anticipated. For these applications, on-site production of pure hydrogen in distributed reactors with integrated palladium membranes could be a solution, as the downscaling of conventional reforming processes based on steam reforming of natural gas is economically not attractive. The hydrogen purification steps, most commonly used is pressure swing adsorption (PSA), are too energy intensive or complex at small scale [2,3]. In literature, also other applications for palladium-membrane reactors are discussed, for example the dehydrogenation of liquid organic hydrogen carriers (LOHC's), such as methylcyclohexane [4]. In these applications, the use of palladium-based membranes represents an efficient approach to in-situ separate hydrogen with ultra-high purity for direct use in PEM fuel cell systems. Furthermore, in equilibrium-limited reactions, continuous hydrogen separation results in advantages with regard to shift in thermodynamic equilibrium conversion and enhancement of reaction kinetics.

Nevertheless, traditional membrane reactors based on porous metallic tubes may not offer the required membrane area per volume of the catalyst to match the H_2 formation rate [4]. As a result, the produced hydrogen is not fully separated from the gas mixture or the activity of the catalyst is not entirely employed. Additionally, tubular arrangements suffer from a rather large distance between catalyst and membrane, which leads to a decrease in partial pressure of hydrogen towards the membrane surface and hence to a reduction of driving force for hydrogen separation (concentration polarization). As a result, the membrane area is not effectively used and larger membrane areas have to be installed.

In this work, a planar microchannel membrane reactor, which is very compact and modular, was designed and constructed.

In a first step, free-standing palladium-based foils of different thicknesses (3 – 61 μm) were leak tight integrated in a microstructured concept module by laser welding. In this concept, the membrane area for hydrogen separation was 1 cm^2 and the potential of the microstructures in decreasing concentration polarization effects was demonstrated [5]. Due to the microchannels, high mass transfer rates and an increased hydrogen partial pressure on the membrane surface compared to conventional systems were achieved [5]. The microstructure served as a good mechanical support for the thin membranes up to temperatures of 400°C and retentate pressures up to 8 bar.

For the application at even higher temperatures and pressures, the palladium foils were supported in a second concept by etched metal sieves coated with a YSZ diffusion barrier layer (DBL) for the prevention of diffusion of metal atoms from the base material into the Pd based membrane. No decrease of the mass transfer rate from the gaseous phase towards the membrane surface was observed due to the additional support and excellent membrane stability was obvious up to 600°C and 11 bar in pure hydrogen permeation experiments (and up to 20 bar retentate pressure at reaction conditions later). The bursting pressure was never reached.

In a second step, different catalysts were prepared for the anticipated reactions methane steam reforming (Rh/Al_2O_3) and water-gas-shift (Pt/CeO_2). The Rhodium-based Al_2O_3 catalysts were inkjet printed in the microstructures and tested in methane steam reforming and water gas shift at ambient pressure in a microstructured reactor in absence of a membrane [6]. Both catalysts showed promising high activity. In methane steam reforming full conversion of methane (98.9%) for a W/F ratio of 5.48 $g_{Cat}h/mol_{CH4}$ and a steam-to-carbon ratio of 4 was obtained. The catalytic performance was modeled with a Langmuir-Hinshelwood mechanism and good agreement with experimental results was achieved.

In a third step, the performance of the modular microstructured concept modules with integrated membranes was evaluated in methane steam reforming and water-gas-shift. Due to the thin catalyst layers a high membrane area-to-catalyst volume was achieved. The microstructured concept module showed promising stability at reaction conditions (temperatures between 623 and 773 K) and high amounts of CO (up to 32 vol. %) did not affect the hydrogen permeability of the Pd membrane.

Based on these experiments, a modular microstructured reactor was designed and built. The electrically heated microstructured membrane reactor contains a section for pre-reforming of methane and a section for catalytic reaction including in-situ hydrogen separation. The results of the ongoing experiments will be presented.

Additional benefits of a microstructured membrane reactor concept will be the heat integration. An outlook on these activities will be given.

References

[1] J.R. Rostrup-Nielsen, "Large-Scale Hydrogen Production", CATTECH, 6 (2002), 150 – 159.

[2] J. Tong, "Pure hydrogen production by methane steam reforming with hydrogen permeable membrane reactor", Cat. Today 111 (2006) 147 – 152.

[3] H. Kurokawa, Y. Shirasaki, I. Yasuda, "Energy-Efficient Distributed Carbon Capture in hydrogen Production from Natural Gas", Energy Procedia 4 (2011) 674 – 680.

[4] R. Dittmeyer. V. Höllein, K. Daub, "Membrane reactors for hydrogenation and dehydrogenation processes based on supported palladium", J. Mol. Catal. A 173 (2001) 135 – 184.

[5] T. Boeltken, M. Belimov, P. Pfeifer, T.A. Peters, R. Bredesen, R. Dittmeyer, "Fabrication and testing of

a planar and microstructured concept module with integrated palladium membranes" Chem. Eng. Process. 67 (2013) 134 – 147.

[6] S. Lee, T. Boeltken, A. Mogalicherla, U. Gerhards, P. Pfeifer, R. Dittmeyer, "Inkjet printing of porous nanoparticle-based catalyst layers in microchannel reactors", Appl. Catal. A (submitted).

(487g) Comparison Of Several Glycerol Reforming Methods For Hydrogen and Syngas Production Using Thermodynamic Analysis
Guirardello, R., UNICAMP
Freitas, A. C. D., University of Campinas - UNICAMP
Originally presented on: 11/6/2013 14:25:00 - 14:43:00

Glycerol, which is obtained as byproduct of biodiesel production, represents an expressive candidate source for hydrogen and syngas production that is renewable. Even with different industrial applications have already been implemented the developments of new technologies for the recovery of glycerol is essential. Among the most promising technologies is the conversion of glycerol into high value added gases, like as hydrogen and syngas using different reforming processes. Among these processes are the steam reforming, partial oxidation autothermal reforming, dry reforming and the supercritical water gasification. All these processes exhibit unique characteristics and the study of the better operational conditions is necessary. This paper focuses on the comparison of different glycerol reforming technologies aimed to hydrogen and syngas production. The reactions of steam reforming, partial oxidation, autothermal reforming, dry reforming and supercritical water gasification were analyzed. For this, the Gibbs energy minimization method was used in combination with the virial equation of state. The validation of the model were made comparing the simulations of the proposed model and results from the literature, either simulated or experimental data, for the different analyzed reactions. The GAMS® 23.1 software and the CONOPT solver were used in the resolution of the proposed problems. The effects of modifications in the operational temperature, operational pressure and reactants composition were analyzed with regard to composition of the products. The effect of coke formation was discussed too. Generally, at higher temperatures and lower pressures resulted in higher hydrogen and syngas production. All cases demonstrated to be feasible for use in the production of hydrogen or syngas. The model showed good predictive ability and low computational time (close to 1 s) to perform the predictions of the combined chemical and phase equilibrium in all systems analyzed here.

(487h) Thermodynamic Analysis of Methane Reforming With CO2, CO2+H2O, CO2+O2 and CO2+Air for Hydrogen and Synthesis Gas Production

Freitas, A. C. D., University of Campinas - UNICAMP
Guirardello, R., UNICAMP
Originally presented on: 11/6/2013 14:43:00 - 15:01:00

Dry reforming of methane for the production of hydrogen and syngas has attracted much attention from both industrial and environmental aspects. The major problem encountered in the application of this process is the rapid deactivation of the catalyst, which is mainly due to coke accumulation and sintering of the catalyst. As an alternative to reduce the coke formation, the combination of CO_2 reforming with steam reforming and oxidative reforming has been proposed in literature. The understanding of the thermodynamic behavior of these systems is important to determine the most favorable reaction conditions. In this context, the main objective of this work is performing a thermodynamic evaluation of methane reforming with CO_2, CO_2+H_2O, CO_2+O_2 and CO_2+air. These evaluations were carried out by Gibbs energy minimization, in conditions of constant pressure and temperature, and entropy maximization, at constant pressure and enthalpy, methods to determine the equilibrium compositions and equilibrium temperatures, respectively. Both cases were treated as optimization problems (using non-linear programming formulation), satisfying the restrictions imposed by atom balance and non-negativity of number of moles. The GAMS® 23.1 software and the CONOPT solver were used in the resolution of the proposed problems. All calculations performed presented a low computational time (less than 1 second). The calculated results were compared with previously published experimental and simulated data with a good agreement between them for all systems. The H_2 and syngas production were favored at high pressure and low temperature conditions. The addition of H_2O, O_2 and air proved to be an effective way to reduce the coke formation in the systems. The CO2 reforming presented endothermic behavior, but the addition of O_2 and air reduced this trend and in some conditions autothermal behavior was observed.

(542a) Desulfurization for Fuel Cells

Gangwal, S., Southern Research Institute
Originally presented on: 11/6/2013 15:15:00 - 15:40:00

Fuel cells can efficiently produce electricity using hydrogen and/or synthesis gas (syngas) obtained by reforming/gasification of gaseous, liquid, and solid fuels. Fuels proposed for use in fuel cell processors include, but are not limited to, natural gas, liquefied petroleum gas, landfill gas, digester gas, gasoline, jet fuel, and solid fuels such as coal and biomass. However, these fuels contain sulfur as impurity that must be nearly completely removed to prevent poisoning of the fuel cell anode catalyst with hydrogen sulfide. Desulfurization methods are critically reviewed and analyzed with particular emphasis on desulfurization sorbents for use in compact fuel processors for fuel cells. Desulfurization can be carried out both upstream and downstream of the reformer. Promising sorbents have been developed for direct deep desulfurization of gaseous and liquid fuels. Also, metal oxide-based sorbents have been developed for desulfurization of syngas down to <0.1 ppmv S. However, further R&D is needed to develop high capacity sorbents for direct desulfurization of liquid fuels and for polishing of syngas. R&D is also needed to simplify sorbent preparation and scale up methods, handling and regeneration, and disposal.

(542b) Characterization of the Silver Phase On Supported Ag/TiO2 Desulfurization Adsorbent Materials Via Multiple X-Ray Techniques

Heinzel, J., Naval Sea Systems Command - Philadelphia
Mansour, A., Naval Surface Warfare Center, Carderock Division
Quinlan, R., Naval Surface Warfare Center, Carderock Division
Tatarchuk, B., Auburn University
Originally presented on: 11/6/2013 15:40:00 - 16:00:00

This paper explores dispersed silver oxides on TiO_2 supports, and intends to identify the electronic and atomic structures and characteristics of the silver phase. Various parameters are given as result of characterization by XPS, XRD, XANES and EXAFS. Significant analysis has been performed to understand how the silver structure changes with loading, with the goal to fully understand how the silver phases may effect the selective interaction of heterocyclic sulfur, and how it can be harnessed and optimized to enhance capacity and ultimately process volume.

XPS analysis confirms that AgNO3, which is used as the precursor for preparing the adsorbent, is not present. Ag is mostly present in an oxide form in the near surface region. Linear combination of XANES data for the TiO_2 supported Ag adsorbent materials show that for all loadings (1, 2, 4, 8, 12, and 20 wt% Ag), the XANES data can be analyzed with a combination of metallic Ag and the oxide phase in 2wt% Ag/TiO_2. For low loadings, 4 wt% and below, only the oxide form is present. For higher loadings, 8, 12 and 20 wt%, the Ag chemistry consists of the oxide form and the metallic form. The metallic content increases with Ag loading. At 2 wt% loading, it was verified that the structure of the oxide form varied with the type of the support (TiO_2 vs. SiO_2).

(542c) Compact, In-Line Fuel Sulfur Analyzer for Improved Management of Desulfurizers

Yelvington, P. E., Mainstream Engineering Corporation
Wagner, A. L., Mainstream Engineering Corporation
Amundsen, T. J., Mainstream Engineering Corporation
Originally presented on: 11/6/2013 16:00:00 - 16:20:00

Fuel cell technology has progressed to the point where fuel desulfurization and reforming have become limiting factors. Both solid-oxide and polymer-electrolyte-membrane fuel cells are poisoned by sulfur compounds in transportation fuels. Sulfur removal is particularly challenging for military fuel cell applications because JP-8 jet fuel contains up to 3,000 ppm of sulfur, 200 times the highway diesel limit. In order to more efficiently manage desulfurization processes, improved in-line methods of determining the extent of sulfur contamination in fuel are required. We are developing a compact, in-line analyzer for measuring total sulfur in jet fuel for fuel cell applications. This analyzer is intended to allow less frequent desulfurizer regeneration events, compared to scheduling regenerations based on the assumed worst-case sulfur limit. The measurement technique employs multivariate analysis of infrared absorption spectra, which provides a robust calibration, outlier detection, and identification of fuel type. The sampling system is non-destructive and allows measurement of sulfur directly in the liquid fuel. Laboratory flow experiments with sulfur adsorbent columns are being performed to evaluate the end use of the sulfur analyzer. Several candidate desulfurizer control strategies that leverage the continuous measurement of fuel sulfur concentration are under evaluation. Sulfur analyzer development and testing as part of a desulfurizer control strategy will be presented.

(542d) Development of a Catalytic Microreactor for Aqueous Phase Reforming of Biofeeds

Neira D'Angelo, F.,
Schouten, J. C., Eindhoven University of Technology
Nijhuis, T. A., Eindhoven University of Technology
Originally presented on: 11/6/2013 16:20:00 - 16:40:00

Hydrogen production by means of Aqueous Phase Reforming (APR) is an excellent way of valorization of waste biomass. In APR, a water soluble carbohydrate and liquid water undergo a reforming reaction at mild conditions (ca. 200 °C, 25 bar) in the presence of a reforming catalyst like Pt/Al_2O_3 to produce highly pure H_2 and CO_2. The low temperatures result in suppression of CO production and reduction of the energy input compared to traditional gas phase reforming. The very low CO content of the gas produced, makes this technology excellently suited for combining it with PEM fuel cells, thereby making this an attractive option for power generation at remote locations.

In this work, we are developing a catalytic microreactor system for aqueous phase reforming. H_2 is selectively produced by combining catalysis and micro-reactor engineering. APR of sorbitol was conducted in a washcoated microchannel. A well adherent, stable and uniform 5 μm layer of Pt/AlO(OH) was washcoated on the walls of a 320 μm ID microchannel. Due to enhanced mass transfer properties, undesired hydrogenation reactions could be avoided, thus enhancing the H_2 formation. The selectivity towards H_2 in the microchannel was greatly increased compared to that in a fixed bed reactor (60 vs. 20 % at the same conversion), resulting in an overall increased turnover frequency (461 h^{-1} vs. 173 h^{-1}). According to these results, the use microtechnology for APR would allow for a great reduction in the reformer size, thus being promising for distributed H_2 production.

(542e) Kinetic Study of Two-Phase Reforming of Aqueous Fraction of Bio-Oil Using Ru Supported On Al2O3

Ayalur Chattanathan, S., Auburn University
Adhikari, S., Auburn University
Srinivasan, V., Auburn University
Abdoulmoumine, N., Virginia Polytechnic Institute and State University
Originally presented on: 11/6/2013 16:40:00 - 17:00:00

Kinetic Study of Two-Phase Reforming of Aqueous fraction of Bio-oil using Ru supported on Al_2O_3

Shyamsundar Ayalur Chattanathan, Sushil Adhikari, Vaishnavi Srinivasan, and Nourredine Abdoulmoumine

Department of Biosystems Engineering, Auburn University, AL

Abstract

Increase in energy demands with population growth has resulted in increased focus on renewable sources of energy. One option is to produce hydrogen from renewable resources such as bio-oil. The objective of this study is to determine kinetics of aqueous fraction of bio-oil in two-phase reforming for hydrogen production. Reforming of bio-oil was carried out in a batch-scale reactor at three different temperatures (180, 230, and 280°C) under autogenous pressure. The effects of temperature and bio-oil concentration (5, 10 and 15 vol. % in water) on exit gas composition and yield were investigated with and without catalysts. The highest hydrogen yield was obtained at 280°C for all bio-oil concentrations. It was observed that the H_2 yield and carbon conversion improved with the addition of Ru/Al_2O_3 catalyst. For example, at 280°C 17% of the carbon was converted into gas phase in the presence of catalyst as compared to 6% without catalyst. All the experiments for kinetics study were conducted at 15% aqueous bio-oil concentration and residence time of 4 h. The activation energy for the reaction was found to be 66 kJ/mol (without catalyst) which decreased to 56 kJ/mol in the presence of catalyst. Although H_2 selectivity was higher with catalytic reforming, there was no significant difference with temperature change. For non-catalytic reforming, it increased with increase in temperature. In addition, H_2 selectivity in catalytic reforming increased by 78%, 58%, and 22% at temperatures 180, 230 and 280ºC, respectively implying that catalyst was effective at lower temperatures.

Keywords: Two-phase phase reforming, Bio-oil, Hydrogen selectivity and Carbon conversion.

(542f) Low-Temperature Decomposition of Hydrous Hydrazine Over Ni-Fe-Cu Nano-Catalysts

Manukyan, K. V., University of Notre Dame
Cross, A., University of Notre Dame
Miller, J. T., Argonne National Laboratory
Wolf, E. E., University of Notre Dame
Originally presented on: 11/6/2013 17:00:00 - 17:20:00

Low-temperature Decomposition of Hydrous Hydrazine over Ni-Fe-Cu Nano-catalysts

Khachatur V. Manukyan[1], Allison Cross[1], Jeff Miller[2] and Eduardo E. Wolf [1]*

[1]Department of Chemical & Biomolecular Engineering, University of Notre Dame, Notre Dame, IN, 46556

[2]*Chemical Sciences and Engineering Division, Argonne National Laboratory, Argonne, IL,* 60429

*ewolf@nd.edu

Hydrous hydrazine ($N_2H_4 \cdot H_2O$) is considered a promising liquid hydrogen storage material owing to the high content of hydrogen (8%) and the advantage of CO-free H_2 produced. We employ a simple, surfactant-free liquid-phase reduction of Cu, Ni and Fe salts (e.g. nitrides, chlorides) to produce nanostructured FeNi/Cu materials for hydrous hydrazine decomposition. Synthesis of nanomaterials includes reduction of copper salt by N_2H_4, followed by rapid reduction of iron and nickel salts by $NaBH_4$. Produced nanoparticles were characterized by XRD, BET, SEM/TEM, XPS, XANES/EXAFS techniques and tested for hydrous hydrazine decomposition at temperatures of 40-70°C. It is shown that the microstructures of resulting FeNi/Cu materials depend on synthesis temperature and concentrations of salts in solution. Experiments show that solely copper nanoparticles (10-20 nn), produced in the first stage were not able to decompose hydrous hydrazine. The FeNi alloy formed on Cu particles during the second reduction stage show activity for decomposition of hydrous hydrazine even at 40°C. However, decomposition products contain ammonia, nitrogen and hydrogen. The selectivity to hydrogen is increased to ~100% with growth of temperature to 70°C. The catalytic performance of these materials primarily depends on the structure of NiFe layer, which may be controlled by simple changing of NiFe:Cu ratio. Investigation of catalytic performance for bi- and tri-metallic materials show that NiFe alloy could be responsible for selective decomposition of hydrous hydrazine. However, the main active metal is nickel. Alloying of nickel with iron coupled with favorable dispersion effect of copper nanoparticle remarkably enhances the catalytic conversion and selectivity of hydrogen evolution.

(542g) Kinetic Modelling of Hydrogen Production By Catalytic Ammonia Decomposition

Monzon, A., University of Zaragoza
Armenise, S., University of Zaragoza
Romeo, E., University of Zaragoza
Garcia-Bordeje, E.,
Valverde, J. L.,
Originally presented on: 11/6/2013 17:20:00 - 17:40:00

Hydrogen production by ammonia decomposition is attracting attention recently as single-step hydrogen generation process. Currently, the technical difficulties involved in the use of hydrogen as an energy carrier are mainly related to transportation and storage, which make the incorporation of this technology slower than initially expected. Among the main competitors of hydrogen as an energy carrier, methane/natural gas and methanol offer clear advantages in terms of energy capacity and distribution infrastructure, but suffer from their intrinsic carbon content causing end-user CO_2 emissions, since in-situ capture is not feasible. However, ammonia combines an easy way to store hydrogen and supply carbon-free stream to the fuel cell stack. In addition, ammonia has other desirable characteristics that make it potentially attractive. It is non-flammable and non-explosive, it can be liquefied under mild conditions, and it has a large weight hydrogen fraction (17.65% of the mass of ammonia). In order to developing efficient reactors for ammonia decomposition for fuel cell applications it is necessary to have realistic kinetic models based on the reaction mechanism.

The ultimate goal of this work is to develop consistent kinetic models able to predict the performance of reactors, like the monolithic reactors, where the NH_3 decomposition is almost complete attaining high H_2 concentrations at the exit. Under these conditions, in addition to the high temperatures needed to attain such large conversions, the gas composition change dramatically along the reactor. Several reaction mechanisms have been suggested, but there is still no consensus about the rate determinant step (RDS), or about the most abundant reactive intermediate (MARI). In spite of the large research effort developed over decades into this reaction, there is no general agreement about which step(s) are the RDSs, or even whether there is a single RDS, and which surface species are kinetically relevant.

The kinetic models developed in this contribution are based on the Langmuir hypothesis considering that all the adsorbed species can be kinetically relevant, that the slow step or steps can be partially reversible, and that the surface can be considered as energetically uniform, i.e. ideal. The results obtained here indicate that the variable values of the kinetic orders and of the apparent activation energies frequently reported in the literature can be direct consequences of the methodology used in the data analysis and can therefore also be explained without considering any change in the controlling step with the reaction temperature or in the hydrogen or ammonia concentration.

(617a) An Investigation of a Stratified Catalyst Bed for Small-Scale Hydrogen Production From Methanol Autothermal Reforming

Richards, N.,
Erickson, P., Hydrogen Production & Utilization Laboratory
Originally presented on: 11/7/2013 8:30:00 - 8:49:00

Introduction

Autothermal reforming (ATR) is a great candidate for use in small-scale hydrogen production given that it is compact, light-weight and has the ability to reform multiple fuels. Unfortunately, as with a steam reforming (SR) system, ATR requires downstream cleanup stages to purify the product stream for use with fuel cell systems.

Research Objectives and Contribution

The overall purpose of this research project is to explore a methanol reforming system using a stratified catalyst bed. This means that by using different catalyst types the reaction will progress according to the catalyst selectivity. The system will include two spatially separated catalysts within a single reactor and will be run autothermally (fuel, steam and oxidizer). The types of catalysts to be used are a standard monolithic noble-metal combustion catalyst, typically used in ATR; along with a commercially available pelletized copper-based water gas shift catalyst used in SR. The ATR catalyst will be upstream and the SR placed immediately behind it. Including a SR section behind the ATR catalyst will allow for a reduction in the limitations experienced by each system.

ATR systems experience mass transfer limitations since the rate of the oxidation step is dependent on the rate of the reactants diffusion to and from the active catalyst sites [1]. This limitation is a critical factor to determine catalyst light-off. Operating below the light-off temperature places the reactions in a kinetically limited region since the combustion process is unable to sustain the reaction; whereas operating above the light-off temperature allows the reaction to be sustained with proper selection of ratios. ATR closely follows an equilibrium reaction [2] and therefore does not achieve high selectivity for hydrogen compared to a SR reaction. Possible ways to overcome this limitation have been investigated by using alternative catalyst compositions [3-5] however the combined effects of geometry and configuration have yet to be explored.

Within the SR catalyst bed, heat transfer limitations are experienced due to the endothermic reaction process. External heat application is typically used to provide the energy needed to activate the catalyst. The catalyst along the reactor walls get heated through conduction. However within the packed bed the random ordering and pellet point-to point contacts cause low conduction rates, therefore convection becomes the dominant mode of heat transfer. Consequently, large temperature gradients are observed near the walls for a SR packed bed reactor which can affect the catalyst performance [6]. In an effort to reduce that particular limitation in SR researchers have explored different techniques. Active [7] and passive [8] methods have been shown to improve the heating of the center of the reactor where the endothermic reaction is occurring. Once the catalyst is active the fuel can be quickly broken apart and reassembled to the desired products based on the catalyst selectivity.

The motivation for the stratified system is that by strategically placing the catalysts in the reactor the reaction will progress according to the catalyst selectivity thereby reducing unwanted intermediates. Essentially this should allow the system to maintain high conversion while ensuring the output gas has a higher concentration of hydrogen than a comparable traditional ATR system. It is also expected to be better equipped to handle transients and have shorter start-up times than a SR system. Additionally, it will be beneficial to small-scale systems since the amount of catalyst can be reduced while maintaining high fuel conversion and hydrogen selectivity. Therefore it should be possible to go to a smaller system in which the fuel breakdown can occur over a small section of noble metal catalyst to be further reformed by a pellet catalyst bed. Overall the findings will contribute to the knowledge of small-scale reforming systems using commercially available catalysts.

References

[1] Pfefferle, L., and Pfefferle, W., 1987, "Catalysis in combustion," Catalysis Reviews Science and Engineering, 29(2-3), pp. 219-267.

[2] Dorr, J., 2004, "Methanol Autothermal Reformation: Oxygen-to-Carbon Ratio and Reaction Progression," M.S., University of California, Davis.

[3] Lindström, B., Agrell, J., and Pettersson, L. J., 2003, "Combined methanol reforming for hydrogen generation over monolithic catalysts," Chemical Engineering Journal, 93(1), pp. 91-101.

[4] Yoon, H., Erickson, P., and Kim, H., 2008, "Lowering the O2/CH3OH ratio in autothermal reforming of methanol by using a reduced copper-based catalyst," International Journal of Hydrogen Energy, 33(22), pp. 6619-6626.

[5] L. Ma, D. L. T., 1996, "Alternative catalyst bed configurations for the autothermic conversion of methane to hydrogen," Applied Catalysis A: General.

[6] Davieau, D. D., 2004, "An Analysis of Space Velocity and Aspect Ratio Parameters in Steam-Reforming Hydrogen Production Reactors," M.S., University of California, Davis.

[7] Erickson, P. A., 2002, "Enhancing the steam-reforming process with acoustics: An investigation for fuel cell vehicle applications," Ph.D. 3065931, University of Florida, United States -- Florida.

[8] Liao, C., 2008, "Hydrogen production enhancement and the effect of passive mixing using flow disturbers in a steam-reforming reactor," Ph.D. 3329636, University of California, Davis, United States -- California.

(617b) In Situ Hydrogen Utilization in An Internal Reforming Methanol Fuel Cell

Avgouropoulos, G., University of Patras
Neophytides, S., FORTH/ ICEHT
Originally presented on: 11/7/2013 8:49:00 - 9:08:00

Though H_2 is being considered as the optimum fuel for PEM fuel cells, a number of reasons render its use difficult, especially in mobile or portable applications. In addition, the lack of a worldwide distribution network with respect to well-established fossil fuel network favors the transition to the real hydrogen economy via a near term solution of on-site (stationary) or on-board (mobile) H_2 production from reforming of various liquid fuels such as gasoline, natural gas, alcohols, etc. The complexity of the balance of plant of a fuel cell-fuel processor unit challenges the design and development of compact and user friendly fuel cell. Depending on the starting fuel different configurations may be applied. The ideal system would be that of a pure H_2-fuelled PEMFC. However, such a system is accompanied with several drawbacks. On the other hand, a voluminous fuel processor with heat exchangers and complex BoP components is required in the case of an alternative liquid fuel as hydrogen carrier. However, the step of enrichment of the hydrogen-stream (i.e. WGS reactor) can be avoided in the case of using methanol as a starting fuel. The direct and efficient processing of methanol has been considered as a potential alternative due to the favorable properties of methanol. The desired power system can be fueled with a renewable, lightweight fuel, such as methanol and will provide uninterrupted power for extended periods at a considerable weight advantage over other standard fuel cell or battery solutions. Future emerging applications include both military and commercial applications such as portable power supply in fires, earthquakes and more, mobile computing, electric/electronic power supply for soldiers, broadband communications, remote monitoring devices.

Recent advances in the design and development of materials, such as polymer electrolyte membranes (e.g. ADVENT TPS® high-temperature polymer electrolyte), electrocatalysts and methanol reforming catalysts, allow the operation of PEMFCs at temperatures in the range 200-220°C. There are several technological and commercial reasons for operating PEMFC at high temperatures:

- Rates of electrochemical kinetics are enhanced
- Water management and cooling is simplified
- Useful waste heat can be recovered
- Lower quality reformed hydrogen containing up to 2% CO may be used as the fuel.

An attractive configuration which would take advantage of the HT-PEMFC features would be the incorporation of fuel reformer inside the fuel cell stack. This requires the development of highly active and selective methanol reforming catalysts, which will be able to operate at temperatures as low as possible with minimal CO formation. Taking advantage of the most recent achievements on the development of reformed hydrogen fuel cells, we propose a novel portable power system based on the incorporation of a methanol reforming catalyst into the fuel cell stack either in separate compartments or directly adjoined with the anode electrodes of the stack, so that methanol reforming takes place inside the fuel cell stack (Internal Reforming).

The proposed fuel cell comprises a high-temperature, membrane electrode assembly (HT-MEA), being able to operate at temperatures of 200-220°C and novel reforming catalysts based on copper – manganese spinel oxides and/or PdZn-based catalysts supported on copper foam or carbon paper, that showed comparative activity at 200-220 °C with commercial $Cu/ZnO/Al_2O_3$ catalysts. The MEA has been developed by ADVENT technologies and the polymer electrolyte is based on crosslinked aromatic polyethers with pyridine units. The MEA has proven operation on a long term basis at 210°C with minimal H_3PO_4 loss for 500 h. The anode is directly adjoined with a methanol reforming catalyst, which will provide the required concentration of H_2. The reformer can be placed in the stack in two different ways: (i) inside the anode compartment adjacent to the Pt-based electrocatalyst, (ii) in between the bipolar plates.

The reforming catalyst is deposited on the surface of conductive structures (i.e. metallic foams or carbon papers) and could be coated by a carbon paste, which efficiently conducts the current out of the MEA to the current collector. If the reforming catalytic bed has a monolithic structure or it is conductive itself it can also operate as a current collector. Finally, current collectors on both sides are directly adjoined with bipolar plates (stainless steel, graphite or graphite composites) that surround the unit cell. The Internal Reforming Methanol Fuel Cell can be supplied with a methanol fuel (mixed with water in appropriate ratios), which is catalytically steam reformed to a H_2-rich gas mixture (it also contains carbon dioxide and water), which together with air supplied on the cathode side drive the electrocatalytic operation of HT-MEA. The outlet stream of the fuel cell contains water and carbon dioxide. The overall design allows for efficient heat management, since the “waste” heat produced by the fuel cell is in-situ utilized to drive the endothermic reforming reaction.

A single-cell of a high-temperature, polymer electrolyte fuel cell incorporating ADVENT TPS® phosphoric acid doped copolymer and a methanol reforming catalyst in the anode has been already constructed and tested at 200°C demonstrating the functionality of the described unit. Specific targets for improvement of the efficiency have been also identified. These are the activity of the reforming catalyst and the thermal stability of the membrane for operation above 200°C. The employed single cell has the following features: (i) triple serpentine flow fields arrangement, (ii) Poco® graphite bipolar plates, (iii) copper current collectors, (iv) electrode area of 8 or 22 cm^2 with a platinum loading of 1.5 mg Pt cm^{-2}, 2 g H_3PO_4/g Pt (v) copolymer TPS® electrolyte membrane, thickness 180 μm, 190 wt. % H_3PO_4 doping level. The membrane was sandwiched between the anodic (Cu-Mn spinel oxide supported on metallic copper foam and placed adjacent to the Pt/C anode electrocatalyst) and Pt/C cathodic gas diffusion electrodes. Vaporized methanol and water mixtures were supplied to the anode compartment, where the reforming catalyst is directly adjoined with the anode electrode. The cell temperature was set at 200°C. Electrocatalytic studies with the proposed configuration confirmed the promoting effect of H^+ pumping through the fuel cell membrane on methanol conversion. Methanol is being reformed inside the anode compartment of the fuel cell at 200°C producing H_2, which is readily oxidized at the anode to produce electricity. The IRMFC (foam of 10 mm thickness) operated efficiently for more than 72 h at 200°C with a current density of 263 mA cm^{-2} at 500 mV, when 20% CH_3OH/30% H_2O/He (anode feed) and pure O_2 (cathode feed) were supplied. Its open circuit voltage was 990 mV. It was interestingly observed that due to H_2 utilization/depletion at the anode the reforming reaction rate was enhanced even up to 20%. The obtained polarization curves and hydrogen production rates at 200°C under fuel cell operation and under OCV conditions showed that significant improvement is achieved by decreasing the reformer thickness from 10 mm to 3 mm. Similar current densities were obtained in both configurations. At 200°C, H_2 utilization/depletion at the anode was facilitated in the case of a foam reformer with 3 mm thickness and resulted in a much higher enhancement of methanol reforming reaction rate as compared with a reformer of 10 mm thickness. In any case, there is enough hydrogen supply for efficient fuel cell operation at 600 mV with 0.2 A cm^{-2} for λ_{H2} (H_2 fed to anode electrode/H_2 reacted) = 1.2. It should be noted though that the methanol conversion obtained with the thinner foam catalyst was lower (75% under fuel cell operation). The activity of current methanol reforming catalysts at the temperature of 200°C is clearly a constraint for further development of the IRMFC concept. There are several advantages of integrating the reformer into the fuel cell:

- Simplification of design. The reformer is placed into the stack, eliminating the need for a separate fuel processor and heat exchangers thus contributing significantly to the stability of the system.
- Increased overall efficiency. In contrast to a conventional PEM fuel cell fed by reformed fuel, the internal reforming fuel cell does not require burning of fuel to drive the reforming reaction because it uses directly the excess heat created by the fuel cell instead. In addition, a heat exchanger is not needed because the reforming catalyst is in direct contact with the heat source.
- Minimization of system weight and volume. Elimination of the WGS and PROX reactors results in a significant decrease in the weight and volume of the whole system.
- The option of in-situ extraction of H_2 from the reforming catalyst via its oxidation by the anode electrocatalyst, enhances the reforming reaction kinetics and results in smaller mass of catalyst (a common kinetic aspect of methanol reforming catalysts is H_2 inhibition).

(617c) Hydrogen Production From Oxidative Steam Reforming of Methanol: Total Oxidation Followed By Steam Reforming of Methanol

Kim, D. H., Kyungpook National University

Originally presented on: 11/7/2013 9:08:00 - 9:27:00

Hydrogen production from oxidative steam reforming of methanol: total oxidation followed by steam reforming of methanol

Dong Hyun Kim, Hyun Chan Lee, Jietae Lee

Department of Chemical Engineering, Kyungpook National University

Daegu, 702-701, Korea

The oxidative methanol steam reforming of methanol (OSRM) has been regarded as a combination of partial oxidation (POM) and steam reforming of methanol (SRM).

POM: $CH_3OH + 0.5\ O_2 \rightarrow CO_2 + 2\ H_2$ (1)

SRM: $CH_3OH + H_2O \rightarrow CO_2 + 3\ H_2$ (2)

By adjusting the O_2 content, the OSRM can be made autothermal, which can make the design of the reformer very simple without heat supply or removal and would be useful in mobile hydrogen production.

Whether POM proceeds according to the equation (2) is not experimentally resolved. At low reaction temperatures where the O_2 conversion was not complete, the products were H_2O and CO_2 rather than H_2 and CO_2 [ref]. Only after the O_2 was completely consumed, the selectivity of H_2 has begun to increase[ref]. In experiments of POM, it was very difficult to keep the catalyst bed isothermal due to the large reaction heat and poor thermal conductivities in the catalyst beds.

In the present study, we developed an isothermal reactor for OSRM using the catalyst supported on aluminum strips. The aluminum plate was anodically oxidized to form a porous alumina layer, on which the active metal components, Cu and Zn, were impregnated. The measured temperature profiles along the bed of 6 cm during OSRM were uniform within ±1°C in the temperature range of 200 – 300 °C. At any reaction temperature in the range and when the O_2 conversion is not complete, the products of OSRM were H_2O and CO_2, showing the total oxidation of methanol in the presence of O_2. It is thus experimentally shown that POM as well as OSRM are methanol combustion until all the O_2 is consumed and then SRM afterward. The reaction rates of the total oxidation were measured and a rate expression was proposed. On the contrary to the implications in the literature, the rate of total oxidation of methanol over the Cu catalyst was observed to be much slower than the rate of SRM in the experimental conditions.

(617d) Design and Control of Stand-Alone Autothermal Reforming Systems Using Methanol

Wu, W., National Cheng Kung University
Wang, C. Y.,

Originally presented on: 11/7/2013 9:27:00 - 9:46:00

Abstract

The conceptual design of compact, stand-alone autothermal reforming systems using methanol is presented. Based on specific conditions of maximum waste heat recovery, the optimization algorithm for maximizing hydrogen yield is obtained. Through the closed-loop control tests for the Aspen Dynamics model, the feasible manipulation by adjusting preheated water flow is established and the stable output regulation of nonlinear systems in the presence of unknown inlet perturbations is guaranteed.

Introduction

Methanol has certain advantages compared to other hydrocarbons due to the low reforming temperature, ease of handling and a relatively high hydrogen-carbon ratio. Similarly, ethanol is also a potentially attractive feedstock because of its availability, nontoxicity and handling safety. From the aspect of kinetics, the reforming processes using different catalysts could affect the hydrogen yield, conversion efficiency and selectivity of undesired products. Hung et al.[1] provided an optimal operating conditions by adjusting H_2O/ethanol and O_2/ethanol ratios to improve catalytic activity. In terms of catalyst utilization, Tang et al.[2] showed that the autothermal reformer has superior performance over steam reformer. The stand-alone fuel cell power system associated with fuel processing units for stationary and potentially mobile applications was recently investigated. Wang and Wang[9] showed that the autothermal reforming system is optimized through thermodynamic equilibrium and exergetic analyses. Stamps and Gatzke[3] showed that a PEMFC stack coupled to methanol reformer with aid of a specific auxiliary power could effectively meet desired power targets. Degliuomini et al.[4]showed the dynamic manipulation and control of a heat integrated PEMFC power system connected to ethanol fueled processing units.

In this work, stand-alone autothermal reforming systems using methanol is evaluated according to hydrogen yield, cost of hydrogen production and energy consumption. As for reducing fuel consumption as well as increasing heat recovery capability, the optimization algorithm for maximizing hydrogen yield of autothermal MeOH-to-H_2 processor subject to the condition of maximum waste heat recovery is performed. Moreover, the dynamic manipulation and control implementation is deployed in Aspen Dynamics environment. According to the closed-loop autotune variation (ATV) tests of the whole dynamics system in the face of unknown inlet perturbations, the PID control method can ensure the no offset and stable output regulation performance. Autothermal MeOH-to-H_2processor

Autothermal alcohols-to-H_2processor

In Fig. 1, a so-called autothermal alcohols-to-H_2 processor is considered as another stand-alone hydrogen production process, where an autothermal reformer (ATR) is an adiabatic plug flow reactor. It is noted that the flue gas from the burner is directly used to adjust the inlet temperature of the ATR, . Regarding the heat recovery design for autothermal alcohols-to-hydrogen processors in Fig. 1, the water flow is preheated by the outlet stream of the WGS reactor using a heat exchanger (E-102), and the feed flow is preheated by the high-temperature flue gas from the burner using anther heat exchanger (E-101). Notably, the heating jacket is removed and the inlet temperature of the ATR, $T_{ATR,in}$, can be adjusted.

Fig. 1 Autothermal alcohols-to-H_2processor

Optimization

To address the maximum waste heat recovery, the following optimization algorithm for maximizing $T_{ATR,in}$. Assumed that $T_{ATR,in}$ is adjusted by changing the heat capacity of exchanger (UA), and UA_urepresents an ultimate bound of UA. If the maximum inlet temperature of the ATR is achieved, i.e. , then the corresponding would be close to the minimum value. Furthermore, the optimization algorithm for maximizing the hydrogen yield. Fig. 2 shows that the maximum hydrogen yield is obtained while the feed ratio (S/C) and are specified at 0.75 and 833.9K, respectively.

Figure 2 Sensitivity analysis

Based on above optimization algorithms, the steady-state simulation under optimal operating conditions is shown in Fig. 3. Notably, it can ensure a lowest flue gas temperature, e.g. $T_{flue,out}$=321K.

Figure 3 Optimization and steady-state simulation

Process control

Aspen Dynamics are employed to model the system dynamics with prescribed steady-state initials. To address a feasible dynamic manipulation, if the initial temperature of $T_{ATR,in}$is 799.4 K and the corresponding UA of heat exchanger (E-101) is fixed at 463.2 K/W, then the open-loop tests of the system with respect to step changes of water flow are shown in Fig. 4.

Figure 4 Open-loop tests of system with step changes of water flowrate

It shows that the manipulation of water flow not only dominates the hydrogen yield but also it affects the waste heat recovery capability. Moreover, a closed-loop ATV test is used to determine the ultimate period and gain when the hydrogen production rate and water flow rate are treated as the controlled output and manipulated input, respectively. The default value of the relay output amplitude is 5%, which is usually good. Fig. 5 shows that ATV test is finished after several (4–6) cycles. Fig. 6 shows that the PID controller settings are determined by Ziegler–Nichols tuning rule.

Figure 5 Closed-loop ATV test

Figure 6 PID controller settings

To check the disturbance rejection capabilities, Fig. 7 shows that the stable and no offset output regulation is achieved while step changes of inlet temperature of water flow appear.

Figure 7 perturbations in the inlet temperature of water flow

By steady state situation, the higher inlet temperature of water flow can induce the higher $T_{ATR,in}$ but $T_{flue,out}$ is not affected.

Conclusions

With respect to hydrogen yield, cost of hydrogen production and energy consumption, the simulation shows that the autothermal MeOH-to-H_2 processor can effectively maximize hydrogen yield as well as minimize heat loss. To address feasible dynamic manipulation, the PID control design using closed-loop ATV tests is employed. Although the autothermal MeOH-to-H_2processor is a typically nonlinear system due to complex kinetics and specific heat recovery conditions, the single-input and single-output (SISO) control structure is successfully established by Aspen Dynamics. Simulation shows that the disturbance rejection by manipulating a preheated water flow can be guaranteed.

References

(1) Hung, C. C.; Chen, S. L.; Liao, Y. K.; Chen, C. H.; Wang, J. H. Oxidative steam reforming of ethanol for hydrogen production on M/Al_2O_3. Int. J. Hydrogen Energy, 2012, 37, 4955.

(2) Tang, H.-Y.; Erickson, P.; Yoon, H. C.; Liao, C.-H. Comparison of steam and autothermal reforming of methanol using a packed-bed low-cost copper catalyst. Int. J. Hydrogen Energy, 2009, 34, 7656.

(3) Stamps, A. T.; Gatzke, E. P. Dynamic modeling of a methanol reformer—PEMFC stack system for analysis and design. J. Power Source, 2006, 161, 356.

(4) Degliuomini, L. N.; Biset, S.; Luppi, P.; Basualdo, M. S. A rigorous computational model for hydrogen production from bio-ethanol to feed a fuel cell stack. Int. J. Hydrogen Energy, 2012, 37, 3108.

(5) Amphlett, J. C.; Creber, K. A. M.; Davis, J. M.; Mann, R. F.; Peppley, B. A.; Stokes, D. M. Hydrogen production by steam reforming of methanol for polymer electrolyte fuel cells. Int. J. Hydrogen Energy 1994, 19, 131.

(6) Choi, Y., Stenger, H. G. Water gas shift reaction kinetics and reactor modeling for fuel cell grade hydrogen. J. Power Sources 2003, 124, 432.

(617e) From Mechanistic to Kinetic Analyses of Ethanol Steam Reforming Over Ir/CeO2 Catalyst

Wang, F., Laboratory of Catalysis, Dalian Institute of Chemical Physics, Chinese Academy of Sciences, Dalian
Cai, W., State Key Laboratory of Catalysis, Dalian Institute of Chemical Physics, Chinese Academy of Sciences, Dalian
Provendier, H.,
Descorme, C.,
Schuurman, Y., Institut de Recherches sur la Catalyse et l'Environnement de Lyon (IRCELYON, UMR 5256, CNRS; Universite© Claude Bernard Lyon 1
Shen, W., State Key Laboratory of Catalysis, Dalian Institute of Chemical Physics, Chinese Academy of Sciences, Dalian
Mirodatos, C., UniversitÃ© Lyon 1, CNRS, UMR 5256, IRCELYON, Institut de recherches sur la catalyse et l'environnement de Lyon

Originally presented on: 11/7/2013 9:46:00 - 10:05:00

From mechanistic to kinetic analyses of ethanol steam reforming over Ir/CeO_2 catalyst

Fagen Wang[a,b], Weijie Cai[a,b], Helène Provendier[b], Claude Descorme[b], Yves Schuurman[b], Wenjie Shen[a], Claude Mirodatos[b*]

[a] *State Key Laboratory of Catalysis, Dalian Institute of Chemical Physics, Chinese Academy of Sciences, Dalian 116023, China*

[b] *Institut de Recherches sur la Catalyse et l'Environnement de Lyon (IRCELYON, UMR 5256, CNRS; Universite© Claude Bernard Lyon 1), 2 Avenue Albert Einstein, 69626 Villeurbanne Cedex, France.*

Despite a large amount of papers dedicated to the ethanol steam reforming (ESR) targeting the production of hydrogen for downstream uses like PEMFC, relatively few ones report a kinetic analysis of the reaction, based on a mechanistic and ageing study for monitoring the kinetic modeling process. To that end, this work presents a combined approach between i) the identification of active sites and their potential ageing, ii) the critical mechanistic steps and iii) a kinetic analysis, accounting for the main mechanistic features.

From thorough investigations carried out over a series of Ir/CeO_2 catalysts [1,2], we demonstrated that the ethanol reforming proceeds essentially via a bi-functional way, involving at least two types of active sites (metal and support) covered with reacting intermediates considered as precursors of the reaction products.

Thus, ethoxy species formed from the reaction of ethanol with the ceria surface, further oxidized into acetate adspecies accumulate on the ceria grains while CH_x and carbonyl pools build up on the Ir particles. The transfer of the C_2adspecies from ceria towards the metal particles, which corresponds to the C-C bond breaking, is strongly monitored by the state of the ceria surface and of the interface between metal and ceria. Again these two parameters are found to be related to the structure (e.g., the density of surface defects on the ceria and the size of the metal particles) and the texture of the catalyst (as deduced from measurements of microporosity and BET surface of the ceria).

Another feature which makes difficult any kinetic analysis of ESR is the occurrence of catalyst ageing, which may render questionable kinetic data collected without accounting for these dynamic processes. Thus, various causes of deactivation were identified, depending on reaction temperature and time on stream. The initial, fast but rather limited deactivation process was ascribed essentially to a loss of ceria surface (smoothing by loss of micro-porosity and/or roughness in the presence of steam), coinciding with an active phase build-up formed by a monolayer of carbonaceous reacting intermediates. In addition, a progressive and long-term deactivation was found to superimpose, originating from structural changes at the ceria/Ir interface, linked to the Ir particles sintering and the ceria restructuring. The continuous build-up of an encapsulating layer of carbon at moderate temperature, coming from C_2 intermediate polymerization (essentially related to the side reaction of ethanol dehydration into ethylene), was found not to contribute significantly to the catalyst deactivation, at least under the operating conditions investigated in this study. This rather stable graphite like layer formed progressively on stream could be suppressed by simple catalyst reoxidation from time to time to avoid potential diffusion limitations.

All these key statements offered us strong guidelines for designing an advanced kinetic modeling of this complex process.

A first conventional kinetic study of ethanol steam reforming over Ir/CeO_2 catalyst was carried out by analyzing the changes in ethanol conversion and selectivity under steam reforming conditions, by varying the main operating parameters (temperature, water/ethanol molar ratio and partial pressure of products). The apparent activation energy of the reaction was measured to be ca 58 kJ/mol, in line with values reported in the literature [3]. The apparent reaction orders of ethanol and water were estimated to be 0.6 and 0.5, respectively, which might indicate in a first approximation that the corresponding adsorption steps are not determining, but participate to the overall rate of conversion via the active ethoxy species and hydroxyl groups. The ethanol conversion was found inhibited by the addition of the main gaseous products in the reaction feed (negative apparent orders, equal to – 0.9, -0.4, -0.4 and -0.3 for H_2, CO_2, CO, and CH_4, respectively), which were consistent with some features of the reaction mechanism. The CO_2 addition would inhibit the ethoxy and acetate migration from ceria to Ir particles upon increasing the carbonate concentration on the ceria. The CO addition would increase the carbonyl concentration at the Ir-ceria interface, thus inhibiting the decomposition of acetate and methyl fragments into CO. The CH_4addition would favor its steam reforming at the expenses of the decomposition of acetate species. As for the detrimental effect of hydrogen addition, a partial reduction of the ceria surface might inhibit key reaction steps like the oxidation of ethoxy to acetate species.

A more advanced kinetic analysis, based on a microkinetic approach, was then carried out by assuming two distinct types of active sites, while the conventional approach based on the Langmuir-Hinshelwood-Hougen-Watson theory generally considers only one type of active site (see for instance [3] or [4]). This analysis will be presented in details during the conference (calculations are still under process), showing the discrimination of various models based on the above mechanistic assumptions. The selected model is considered to be robust enough to explore advanced engineering solutions for the hydrogen production from ethanol reforming, such as using micro-reactors as a tool for process intensification, as described in [5].

References:

1- Ageing analysis of a model Ir/CeO2 catalyst in ethanol steam reforming. F. Wang, W. Cai, Tana, H. Provendier , Y. Schuurman, C. Descorme, C. Mirodatos·, W. Shen, *Appl. Catal. B*, **2012**,*125,*546.

2- Oxidative steam reforming of ethanol over Ir/CeO2 catalysts: A structure sensitivity analysis. W. Cai, F. Wang, C. Daniel, A. C. van Veen, Y. Schuurman, C Descorme, H. Provendier, W. Shen, C. Mirodatos, *J. Catal.* **2012**, *286,*137.

3- Ethanol Steam Reforming over Rh(1%)$MgAl_2O_4/Al_2O_3$: A Kinetic Study, C. Graschinsky,. M. Laborde, N. Amadeo, A. Le Valant, N. Bion, F. Epron, D. Duprez, *Ind. Eng. Chem. Res.* **2010**, *49,* 12383

4- Experimental, kinetic and 2-D reactor modeling for simulation of the production of hydrogen by the catalytic reforming of concentrated crude ethanol (CRCCE) over a Ni-based commercial catalyst in a packed-bed tubular reactor. E. Akpan, A. Akande, A. Aboudhier, H. Ibrahim, R. Idem, *Chem. Eng. Sci.* **2007**, *62*, 3112.

5- Hydrogen production from ethanol steam reforming in a micro-channel reactor. W. Cai, F. Wang, A. van Veen, C. Descorme, Y. Schuurman, W. Shen, C. Mirodatos, *Int. J. Hydrogen Energy*, **2010**, *35,* 1153

(617f) Design and Demonstration of an Ethanol Fuel Processor for HT-PEM Fuel Cell Applications

Gardemann, U., Zentrum fÃ¼r BrennstoffzellenTechnik, ZBT GmbH
Steffen, M.,
Heinzel, A., Univerity of Duisburg-Essen
Originally presented on: 11/7/2013 10:05:00 - 10:24:00

Introduction and Aim

The aim of this work was the design and demonstration of a compact ethanol fuel processor for small scale fuel cell systems. Promising applications are the independent power supply for stationary off grid applications and backup power as well as leisure and security markets.

Ethanol serves as CO_2neutral fuel as it can be produced from renewable sources. Ethanol furthermore shows technical advantages in terms of high energy density, portability and storage capability so that it is a promising fuel for the mentioned stationary and mobile small power output fuel cell applications.

The fuel cell type to be considered within this development is a high temperature PEM-fuel cell (HT-PEMFC), which shows good CO-tolerance due to the high operation temperature of around 160 °C and therefore requires less complex fuel processing. Further advantages of the higher operation temperature are the potential of effective stack cooling even with media like ambient air and a less complex cathode air supply as no humidification is required [1].

The rated electrical power output of the system is set to be P_{el} = 200 – 500 W, which means that the fuel processor has an nominal hydrogen output of $P_{H2,th}$ = 600 – 1500 W. Other requirements for the fuel processor are fast start-up, high robustness with stable operation during planned or undesired parameter changes (e.g. ambient temperature, components and catalyst wearing …), the potential of low cost manufacturing and a low pressure drop to keep the electrical power consumption small. Issues as overall efficiency, maximal heat integration and dynamic operation are not focussed at the beginning of this development.

The autothermal (ATR) or oxidative steam reforming is the preferred reforming principle as it shows a high potential to meet most of the requirements especially robust operation and fast start-up. In addition ATR operation strongly reduces the risk of coke formation and the energy demand for the reforming process [2] [3]. The heat balance can actively be adjusted by varying the oxygen feed.

Main components of the fuel processor are the reformer reactor, the shift-converter and a catalytic burner combined with heat exchangers in order to meet the temperature demands of the different reaction stages.

Special attention turns on the evaporation of the liquid media ethanol and water for the reformer and burner, respectively.

Approach and Results of the Work

In order to evaluate the most appropriate parameter set-up for the fuel processor a theoretical analysis of the complete fuel cell system has been carried out with Aspen Plus®. As a result of sensitivity analyses a basic operation point was defined with a good compromise between high hydrogen yield, heat balance and low CO-concentration. The main operation parameters namely steam to carbon ratio (S/C), oxygen to carbon ratio (O_2/C), reforming temperature (T_{Ref}) and shift temperature (T_{Shift}) were set to 2.75, 0.40, 700 °C and 300 °C, respectively.

Furthermore a heat integration analysis by means of the pinch point method has been performed. As result the heat demand for the evaporation and superheating of the reactants can completely be met with the energy content of the anodic off gas assuming a fuel utilization of 0.70 in the fuel cell. No further heat integration is necessary to keep the process running. Therefore the degree of freedom can be kept high, e.g. the shift temperature can be adapted without influencing other streams or parameters.

Due to the presence of oxygen in the reactants precious metal catalysts are compulsory. Furthermore for the autothermal reformer these catalysts have the potential to suppress the formation of the undesired by-product ethylene which serves as a coke precursor [4].

As one experimental part of the work a catalyst screening of reforming catalysts has been carried out at the beginning. Two commercially available catalyst formulations deposited on ceramic monoliths show similar performance. A slight advantage of the chosen catalyst is given concerning high hydrogen yield at high gas hourly space velocities (GHSV) of about 50000 1/h.

The choice of shift and burner catalysts as well as the design of these reactors was based on former experimental work at the ZBT [5] [6]. Good experiences exist with commercial precious metal catalysts which are deposited on ceramic monoliths for those stages as well. Possible GHSV in the range of 20000 1/h up to 50000 1/h resulted in compact shift-converter and burner reactors.

The developed burner contains an electrically heated zone equipped with a metal fibre element upstream of the catalyst for the ethanol to be vaporized homogeneously and the catalyst to be heated during start-up. About five minutes after cold start complete evaporation of the inlet ethanol air mixture and full conversion in the burner catalyst can be reached. For this start-up process energy of about 25 Wh is required. The switchover to anodic off gas (AOG) can easily be performed without interruption. Stable combustion could be achieved for both fuel mixtures.

The evaporation and superheating of the reactants ethanol, water and air takes place in an integrated unit supplied with heat from the catalytic burner. A metal foam structure serves as evaporation support and provides independence of the system position. The integrated unit had to be designed precisely in order to avoid thermal decomposition of ethanol. A proof of concept could be demonstrated operated with the burner. Outlet temperatures up to 350 °C can be reached. The pressure surges are in the range of several millibars which indicates homogeneous evaporation.

Finally all units were assembled to a complete fuel processor, which could be qualified with various operation parameter set-ups. The theoretical defined basic operation point could be confirmed as suitable. The overall start-up time to receive a reformat with appropriate quality to feed a HT-PEMFC (x_{CO} < 2 %) takes around 30 minutes. At steady state operation the hydrogen power output is around 900 W with $H_2$and

CO fractions of 41.2 % and 1.5 %, respectively.

Literature

[1] Bandlamudi, G.: *Systematic characterization of HT PEMFCs Containing PBI/H3PO4 systems. Thermodynamic analysis and Experimental investigations*. PhD Thesis, University of Duisburg-Essen, Logos, Berlin, 2011, ISBN 978-3-8325-2962-8

[2] Rabenstein, G.; Hacker, V.: *Hydrogen for fuel cells from ethanol by steam-reforming, partial-oxidation and combined auto-thermal reforming: A thermodynamic analysis*. Journal of Power Sources 185 (2008), p. 1293 – 1304

[3] Cavallaro, S.; Chiodo, V.; Vita, A.; Freni, S.: *Hydrogen production by auto-thermal reforming of ethanol on Rh/Al2O3 catalyst*. Journal of Power Sources 123 (2003), p. 10 - 16

[4] Shekhawat, D.; Spivey, J.J; Berry, D.A.: *Fuel Cells: Technologies for fuel processing*, Elsevier B.V. 2011

[5] Spitta, C.; Dokupil, M.; Mathiak, J.: *Compact propane fuel processor for auxiliary power unit application*. Journal of Power Sources 157 (2006), S. 909 – 913

[6] Spitta, C.; Dokupil, M.; Mathiak, J.; Heinzel, A.: *Coupling of a small scale hydrogen generator and a PEM fuel cell*. Fuel Cells 7, 3 (2007), S. 197-203

(617g) Effect of Alumina On Oxidative Steam Reforming of Ethanol for Minimization of CO Formation and Maximization of H2 Selectivity

Sajnani, M.,
Pant, K. K., Indian Institute of Technology, Delhi
Originally presented on: 11/7/2013 10:24:00 - 10:43:00

As conventional fuels deplete rapidly, reforming of hydrocarbon feedstocks for hydrogen generation and production of clean energy gains huge importance. Ethanol is a potentially good feedstock for reforming because of its non toxicity and free availability. Therefore developing a new cheap, stable catalyst for H_2 yield maximization and CO minimization during reforming at optimum conditions is of core importance.

The study of oxidative steam reforming of ethanol was performed using Ni/Al_2O_3-CeO_2-ZrO_2 and Ni/CeO_2-ZrO_2 catalyst to investigate the hydrogen production in a micro-reformer. The catalysts were prepared by oxidative co-precipitation followed by wet impregnation method and were characterized using Temperature Programmed Reduction (TPR), X-Ray Diffraction (XRD), and Energy dispersive X-ray (EDX) and BET surface area analysis. BET surface area analysis showed multifold increase in pore volume and area for alumina based catalysts as compared to the other catalyst. Various kinetic parameters like time on stream, temperature, pressure, oxygen/ethanol and ethanol/steam ratios were studied. Effects of addition of alumina and reforming temperature were studied for hydrogen production at water/ethanol/oxygen ratio of 9:1:0.35 in a temperature range of 550° C to 750 ° C. Operating parameters were predicted from thermodynamic studies done prior to the experimental analysis. Effect of contact time was also studied by performing experiments with W/F_{A0} ranging from 3.4 to 13.8 kgcat.h/kgmol[ethanol]. Selectivity trends of H_2, CO, CO_2 and CH_4 were close to the predicted values attained by thermodynamic analysis. As the reforming temperature increases extent of methanation, methane dry reforming and water gas shift reaction decreases leading to increase in CO selectivity and reduction in CO_2 and CH_4 selectivity. Effect of temperature on conversion and yield for both the catalysts was studied and a comparison based on H_2 selectivity, yield and CO selectivity revealed that alumina based catalyst is better in terms of its stability and its performance. Results showed that hydrogen selectivity and conversion for alumina based catalyst was higher where as methane and CO selectivity was lower relatively. Significant yield of H_2 corresponding to (74%) 3.66 moles (out of 4.9-thermodynamic maximum) per mol of ethanol and selectivity (67%) was obtained at optimal conditions for Ni/Al_2O_3-CeO_2-ZrO_2 catalyst.

(617h) From Waste to Gas: Biogas Valorization

Barrio, V. L., Faculty of Engineering (UPV/EHU)
Izquierdo, U., Faculty of Engineering (UPV/EHU)
Requies, J.,
Guemez, M. B.,
Cambra, J. F., Faculty of Engineering (UPV/EHU)
Arias, P. L.,
Originally presented on: 11/7/2013 10:43:00 - 11:02:00

From waste to gas: biogas valorization

V.L. Barrio, U. Izquierdo, J. Requies, M.B. Güemez, J.F. Cambra and P.L. Arias

Faculty of Engineering (UPV/EHU), c/ Alameda Urquijo s/n, 48013 Bilbao, Spain

Introduction

In this work biogas valorization – a renewable resource – for synthesis gas and hydrogen generation through reforming processes is studied using a conventional fixed bed reactor and an advanced microreactor system.

It is time for hydrogen as an energy vector and this opportunity should be seized by doing improvements and innovations in the existing technologies and processes as well as for developing new ones. This renewable hydrogen can also be used for the Power to Gas technology.

The behaviour of two different catalytic systems was studied in this work. On the one hand, Ni-based catalysts and a bimetallic Rh–Ni catalyst supported on magnesia or alumina modified with oxides like CeO_2 and ZrO_2were used comparing conventional and microreactor reaction systems. On the other hand, Ni monometallic and Rh-Ni bimetallic based catalysts were prepared supported on three different Zeolites L.

For all the experiments, a synthetic biogas (molar composition: 60% CH_4 and 40% CO_2) was fed and the catalytic activities were measured in two different experimental facilities: a bench-scale fixed bed reactor system and a microreactor reaction system, at 1073 K and atmospheric pressure. Physicochemical characterization of catalyst samples by ICP-OES, N_2 physisorption, H_2chemisorption, TPR, SEM, TEM, XRD and XPS showed differences in chemical state, metal–support interactions, average crystallite sizes and redox properties of nickel and rhodium metal particles, indicating the importance of the morphological and surface properties of metal phases in driving the reforming activity.

Results & Discussion

Magnesia – Alumina modified catalysts

Several Ni-based catalysts and a bimetallic Rh–Ni catalyst supported on magnesia or alumina modified with oxides like CeO_2 and ZrO_2were used. In this sense, nickel monometallic and rhodium-nickel bimetallic based catalysts were prepared, characterized and tested in dry reforming (DR), steam reforming (SR), oxidative reforming (OR) and tri-reforming (TR) processes for syngas and/or hydrogen generation.

The catalysts which achieved higher activity and stability in the fixed-bed reactor were impregnated in a microreactor to explore possible process intensification. For TR processes, different steam to carbon ratios, S/C, from 1.0 to 3.0, and O_2/CH_4ratios of 0.25 and 0.50 were used.

For the Ni-based catalysts and the bimetallic Rh–Ni catalyst supported on magnesia or alumina modified with oxides like CeO_2 and ZrO_2, high methane and carbon dioxide conversions were reached for DR reaction in the fixed bed reactor system. However, this process requires huge energy supply and carbon filaments growth was observed by SEM. Regarding biogas TR process, different O_2/CH_4 and S/C ratios were tested. For this process, at O_2/CH_4= 0.25 and S/C = 1.0 the highest carbon dioxide conversions and hydrogen yields were reached.

The Rh–Ni/Ce–Al_2O_3 catalyst achieved the highest hydrogen production yield in DR reaction and also in the biogas TR at the following experimental conditions: O_2/CH_4= 0.25 and S/C ratio of 1.0. These results can be due to its very high active metal dispersion as well as the reached small particle size measured by XPS and XRD.

Operating with microreactors similar conversions were achieved being Ni/Ce–Zr–Al_2O_3 catalyst the one reaching the highest hydrogen production yield for biogas TR at O_2/CH_4= 0.25 and S/C = 1.0. Attending to the TOF and productivity values, the catalytic activity measured using microreactors was one order of magnitude higher than the achieved one using fixed bed reactor. In addition, process intensification improved catalysts stability.

Zeolite supported catalysts

For the Ni-based catalysts and the bimetallic Rh–Ni catalyst supported on zeolites, three different Zeolites L were synthesized to be used as catalyst supports. A disc-shape Zeolite L (length around 0.2-0.4 μm and diameter of approximately 0.5 μm) and two Zeolites L with a cylindrical morphology (smaller particle size around 30-60 nm and bigger particle size around 1-3 μm) were tested.

The operation conditions for the experiments carried out with the Zeolites L were chosen taking into account the previous results obtained for the alumina supported catalysts, that correspond to the following: for biogas SR steam to carbon (S/C) of 1.0 and 2.0, for biogas OR the O_2/CH_4 ratio of 0.25 and 0.50, and S/C=1.0 and for TR reactions O_2/CH_4=0.25.

Focusing on the rest of catalysts and according to the morphological difference between the Disc and Cylindrical (30-60 nm) Zeolites L, different activities were measured. For the SR processes, the Disc catalysts were more active at the lowest S/C ratio and the activity decreased when the S/C ratio was increased to 2.0. By the contrary, the Cylindrical (30-60 nm) catalysts are more active in the OR at O/C=0.25 and decreased when the S/C ratio was increased to 0.5.

In all tested processes the Ni and Rh-Ni catalysts based on the Zeolite L (1-3μm) achieved the lowest activity values. This could be associated to the presence of Merlinoite zeolite –detected by XRD- which contributes in around the 25% of the sample.

With the exception of the Ni Disc catalysts for SR at S/C=1.0 and DR processes, the bimetallic catalysts showed much better reforming capacities in the tested processes and conditions. Thus, Rh incorporation increased catalytic activity in all the reforming processes tested. The Rh-Ni catalysts based on Disc and specially the Cylindrical (30-60 nm) Zeolite L seemed to be a very promising catalyst due to its very high metallic dispersion –measured by H_2chemisorption– and excellent activities.

Regarding the processes, for biogas SR and TR processes, H_2/CO ratios near to 2.0 were obtained and this is an appropriate ratio for the Fischer Tropsch synthesis.

Finally, attending to the CO_2 reforming capacity, DR is the most appropriate process for the conversion of almost all the CO_2that makes up the biogas. .

References

1) Winter CJ. *Int. J. Hydrogen Energ.*34 (2009) S1.

2) Rasi S, Veijanen A, Rintala J. *Energy*32 (2007) 1375.

3) Kolbitsch P, Pfeifer C, Hofbauer H. *Fuel*87 (2007) 701.

4) Dehmer, Dagmar. *The Electricity Journal*26(1) (2013) 71.

5) Gahleitner, Gerda. *Int. J. Hydrogen Energ.*38 (2013) 2039.

6) Murphy, Jerry D.; James Browne, Eoin Allen, Cathal Gallagher. Renew. Energ. 55 (2013) 474.

Hydrogen Production By Catalytic Conversion Of Synthetic Fuels

Advanced Fossil Energy Utilization
AIChE Annual Meeting
Opwfn cfs8-3324 .22;33bn .22;26bn

For fuel cell applications, the production of hydrogen by steam reforming of synthetic fuels such as dimethyl ether, methanol, and dimethoxymethane was studied over bifunctional ÑuO-CeO_2/g-Al_2O_3catalyst. The catalyst was characterized by XRD, HR TEM, EDX and FTIR spectroscopy. It was found that the catalyst provided the hydrogen production rate ~15 l $H_2/(g_{cat} \cdot h)$ at complete conversion of dimethyl ether, methanol and dimethoxymethane to hydrogen-rich gas with low CO content in the temperature interval of 300–350°C. Thus, bifunctional catalyst demonstrates high promises for the realization of multifuel processor concept. It enables the production of hydrogen-rich gas from more than one fuel (DME, methanol and DMM) under similar reaction conditions. Moreover, the produced hydrogen-rich gas can be used for direct feeding HT PEM FC and, after CO removal by CO PROX with the same catalyst, LT PEM FC.

(693a) Fischer-Tropsch Synthesis: Matching the Catalyst to the Feedstock and the Reactor

Jacobs, G., University of Kentucky
Pendyala, R. R., University of Kentucky
Ma, W., University of Kentucky
Gnanamani, M. K., University of Kentucky
Davis, B. H., University of Kentucky
Originally presented on: 11/7/2013 12:30:00 - 12:55:00

Only three elements have been found suitable for the Fischer-Tropsch synthesis: iron, cobalt and ruthenium. However, for each of these elements a variety of catalyst formulations are possible. It therefor becomes a problem of mating a catalyst formulation to the properties of the feedstock. While any carbon containing feedstock can be used to generate a syngas, the impurity compounds and levels may vary with each feedstock as will the hydrogen/carbon monoxide ratio that is produced. The adjustment of the hydrogen/carbon monoxide ratio to that needed for synthesis will define the reactors needed as well as the catalyst. In addition, four reactor types have been utilized for Fischer-Tropsch synthesis: fixed- and circulating-fluid bed, fixed bed and slurry reactor. In addition small channel reactors are in the development stage and these offer attractive features. The properties, both physical and chemical, must then be matched to those needed for each reactor type.

(693c) Development of Novel Carbon Nanotubes Supported Catalysts for Fischerâ€"Tropsch and Higher Alcohol Syntheses

Dalai, A. K., University of Saskatchewan
Abbaslou, R., University of Saskatchewan
Trepanier, M., University of Saskatchewan
Surisetty, V. R., University of Saskatchewan
Originally presented on: 11/7/2013 13:13:00 - 13:31:00

Abstract

Fischer-Tropsch (FT) and higher alcohol synthesis (HAS) processes are important for the conversion of synthesis gas (syngas) to liquid fuels. These fuels are free from sulfur and nitrogen compounds and are thus quite clean with high octane and cetane numbers. In this presentation, the feasibility of using new nanocatalyst formulations for FT and HAS are discussed with a focus on the conversion of syngas to green diesel and alcohols. For FT synthesis, the effectiveness of both unpromoted and promoted Fe and Co on carbon nanotubes (CNTs) is examined. The effects of pore diameter and structure of iron and iron catalysts supported on CNTs on FT reaction rates and product selectivity are presented. Two types of CNTs with different average pore sizes (12 and 63 nm), however with comparable surface area, were examined. The CNTs were prepared by chemical vapor deposition method, where as the metals were loaded with incipient impregnation method and the catalysts were characterized by ICP, BET, XRD, TPR, SEM and TEM analyses. According to TEM images of iron catalysts supported on narrow pore CNTs (np-CNT) and wide pore CNTs (wp-CNT), a majority (~80%) of the iron oxide particles were deposited inside the nanotubes' pores. These particles on the Fe/wp-CNT (17 nm) were larger than those on Fe/np-CNT sample (11 nm). Also, the degree of reduction and metal dispersion of the Fe/np-CNT catalyst was much higher compared to the other catalyst leading to better FTS in terms of CO conversion and C_5+ selectivity of the np-CNT catalyst. In addition, novel cobalt catalyst supported on CNTs promoted with Co, Ru and K with low deactivation was created, extensively characterized and studied for FT reactions. The data show that the catalyst is quite promising with high catalyst activity and C_5+ selectivity. A series of alkali and metal (Rh and Co) promoted MoS_2 catalysts supported on CNTs were prepared and extensively tested for HAS reactions in a fixed-bed system. Our research has shown that the total alcohol space time yield (STY) using trimetallic catalyst is 0.281 g/g of cat.h at optimum reaction conditions.

(693d) Supported Cesium-Promoted Molybdenum Sulfide Catalyst and Reaction Kinetics for Higher Alcohol Synthesis

Caram, H., Lehigh University
Sahoo, R.,
Herman, R. G., Lehigh University
Acharya, D.,
Originally presented on: 11/7/2013 13:31:00 - 13:49:00

Supported Cesium-promoted Molybdenum Sulfide Catalyst and Reaction Kinetics for Higher Alcohol Synthesis

Hugo S. Caram[1], Ranjan K. Sahoo[1], Richard G. Herman[1], Divyanshu R. Acharya[2]

[1]*Department of Chemical Engineering, Lehigh University, Bethlehem, PA 18015 USA*

[2]*AMCS Corporation, Bedminster, NJ 07921*

Cesium-promoted molybdenum catalyst supported on activated carbon has been prepared and tested in a fixed-bed configuration for higher alcohol synthesis from syngas (H_2+CO). Effect of cesium loading, process temperature, pressure and hydrogen to carbon monoxide ratio on catalyst activity, selectivity and stability were studied. Alcohols productivity of 280 g/kgcat/hr (Ethanol productivity of 100 g/kgcat/hr) with selectivity of 85% w/w was achieved at 315 C, 1200 psig and H2/CO =1. The catalyst was stable for more than 500 hours without any loss of activity and selectivity for alcohols. A kinetic model of Langmuir-Hinshelwood type rate expressions was developed and fitted with experimental data. A Singular Value Decomposition method was use to minimize the stoichiometric description. A Genetic Algorithm optimization tool was used to estimate kinetic parameters of these models. Evaluation of these kinetic models has been done using a structural representation of the catalyst behavior with respect to residence time of reactants in the catalyst bed.

(693e) Elucidation of the Active Phase of K Promoted MoS2 for the Direct Production of Higher Alcohols From Synthesis Gas: An Operando Infrared Study

Santos, V., Delft University of Technology
van der Linden, B., Delf University of Technology
Chojecki, A.,
Budroni, G.,
Corthals, S.,
Shibata, H.,
Meima, G.,
Kapteijn, F., Delft University of Technology
Makkee, M., Delft University of Technology
Gascon, J., Delft University of Technology
Originally presented on: 11/7/2013 13:49:00 - 14:07:00

Elucidation of the active phase of K promoted MoS_2 for the direct production of higher alcohols from synthesis gas: an operando infrared study

Vera P. Santos[1], Bart van der Linden[1], Adam Chojecki[2], Gerolamo Budroni[2], Steven Corthals[2], Hirokazu Shibata[2], Garry R Meima[2], Freek Kapteijn[1], Michiel Makkee[1],
and Jorge Gascon[1]*

[1]Catalysis Engineering/ChemEng., Technical University of Delft, Julianalaan 136-2628 BL Delft, The Netherlands

[2]Dow Benelux B.V., P.O. Box 48-4530 AA Terneuzen, The Netherlands

e-mail: j.gascon@tudelft.nl

Introduction

The increasing demand for renewable octane booster has drawn considerable interest towards the synthesis of mixed alcohols. Although bio-ethanol and to some extent methanol are currently used as gasoline additives with a high octane number (> 100), higher alcohols are preferred because of their lower volatility, higher solubility in hydrocarbons and higher calorific value [1]. Coal, biomass, and natural gas can be converted into higher alcohols via synthesis gas (Fischer-Tropsch synthesis) over several transition metals [2]. Among them, MoS_2 is one of the most promising catalysts, since it shows high resistance to sulphur poisoning and deactivation by coking [3]. In contrast to un-promoted MoS2, that produces mainly hydrocarbons (C1-C6), catalyst based on alkaline promoted MoS2 display a high selectivity towards higher alcohols. Despite extensive investigations on the synthesis of hydrocarbons and alcohols the mechanism over MoS_2 based systems, there is still no agreement on the nature of the active sites and the role of alkali promoters.

Here, we apply *operando* infrared spectroscopy in combination with high throughput (HTP) screening measurements to elucidate the role of potassium on the activation of carbon monoxide on the MoS2 surface and stabilization of reaction intermediates.

Materials and Methods

Unsupported MoS2 was prepared by thermal decomposition of (NH4)2MoS4 in nitrogen at 520 °C. The promotion was carried out by physical mixture of MoS2 with potassium carbonate (6% wt K).

Syngas synthesis (1:1 H2/CO ratio) was performed on a fixed-bed reactor, operated at 90 bar, 310 °C and using a gas hourly space velocity of 10,000 l/l/h..

Operando DRIFTS was carried out at 350 ^{o}C, 30 bar, a space velocity of 7000 l/l/h., and with a ratio CO/H2 of 1, using an *in-situ* high pressure cell and undiluted catalyst samples (no KBr). It was established that the breaking period for the reactor setup at 310 °C and 90 bars will give the same results as that of 350 °C and 30 bar in the DRIFT set-up.

Results and Discussion

The conversion profile and alcohol selectivity as a function of time on stream for some catalysts are shown in Figure 1a. The conditioning of molybdenum sulfide catalysts is characterized by long transient periods (break-in), accompanied by a continuous decrease in activity due to the permanent reconstruction of the active phase. However, the major significant differences in the evolution of the conversion of carbon monoxide and in the product distribution occur during the first 120 hours. For this reason, the *operando* study was focused on this very onset of the break-in period. As demonstrated by the screening studies, the presence of potassium on MoS2 shifts the selectivity towards alcohols (C1-C4 alcohols, 45% selectivity) and decreases the deactivation rate of MoS2, showing that potassium acts simultaneously as a chemical and structural promoter. Figure 1b shows the evolution of hydrocarbons and alcohols as function of time on stream for the promoted MoS2 (K-MoS2) obtained from the *operando* DRIFTS study. A complete match between the HTP experiments and operando DRIFTS was observed. The time resolved *operando* IR spectra illustrate the formation of hydrocarbons and CO2 already from the very beginning of the break in stage. Subsequent analyses of the adsorbed species by infrared spectroscopy reveal that alkoxy species are an important intermediate in the syngas to alcohol route, and that potassium plays a strong role on the stabilization of these species.

The electronic properties of MoS2 also change upon promotion, thereby allowing for a relatively enhanced activation of the CO molecule in respect to the hydrogen-assisted CO activation route over MoS2.

Conclusions

Several infrared *in-situ* works have been reported on alkali promoted MoS2 catalysts for the synthesis of higher alcohols, but most of them deal with supported catalysts. The few IR studies on bulk MoS2 use KBr as diluent, something surprising if the promotion effect of K is to be unraveled. In this work we make a crucial step in resolving the role of alkali promotion on CO activation, stabilization of intermediates, and intrinsic properties of MoS2 by combining *operando* spectroscopy and high-throughput experimentation.

This insight is crucial for the further development of a syngas process for the production of higher alcohols.

References

[1] R. G. Herman, *Catal. Today 55,* 233 *(2000).*

[2] Jacob A. Moulijn, Michiel Makkee, and Annelies van Diepen in: *Chemical Process Technology,* John Wiley & Sons Ltd, England, **2001.**

[3] S. Zaman, K. J. Smith, *Catal. Rev. 54,* 41 *(2012).*

(693f) CO Hydrodeoxygenation On Functionalized Metal-Free CNT Surfaces

Mondal, K., Southern Illinois University
Talapatra, S., Southern Illinois University
Jeffers, M., Southern Illinois University
Halfacre, K., Southern Illinois University
Pokhrel, S., Southern Illinois University
Vinson, H.,
Liang, L.,
Elias, A. L.,
Zhang, X.,
Sumpter, B. G., Oak Ridge National Laboratory
Meunier, V., Rensselaer Polytechnic Institute
Terrones, M.,
Originally presented on: 11/7/2013 14:07:00 - 14:25:00

CO hydro-deoxygenation can lead to products such as hydrocarbons (obtained through Fischer Tropsch Synthesis) and oxygenates (such as alcohols ketones and aldehydes) as well as cyclic and aromatic compounds. The long carbon chain moieties of these organic compounds have significant importance as fuels. Traditionally, late transition metals such as iron and cobalt are use as active catalysts with a number of promoters that affect the reducibility of these metals and on the activity and selectivity. Herein, we report the results of the experimental and theoretical investigations on CO hydro-deoxygenation and subsequent chain growth on metal free CNT surfaces functionalized with C-O, COOH, OH and N. Experimental data showed that these catalysts prepared by air assisted CVD show significant enhancement (1-3 orders higher) in the rate of conversion as compared to a simple Fe-Zn-K catalyst supported on gamma alumina that was produced by precipitation-impregnation method. XPS data show no Fe (catalyst for CNT growth) signature and it was concluded that the Fe, if present, is occluded from the reactants. The experimental data and XPS showed that the reaction rates correlated well with the oxygen containing functional groups on the CNT molecules. The Raman spectra also showed that an increase in the defects improved the conversion rates. The TGA data show that the occluded iron was less than $1/8^{th}$ of the amount of that used for the Fe-Zn-K catalyst. A complete characterization of the experimental conditions was conducted and it was observed that the activity of the CNT catalysts ranged from 200 – 450 ^{o}C. It was observed that the product distribution could be tuned to obtain oxygenates or linear chain hydrocarbons by adjusting the reaction temperature or pressure. DFT calculations conducted independently showed that C=O or OH functional groups are essential for the progress of this reaction. It was calculated that while pristine CNTs could not chemisorb CO or H_2, CO could be chemisorbed on to epoxy or COOH sites. It was also seen that the hydrogen insertion could not occur from the gas phase and the mechanism required the presence of OH groups which could be present on the as produced material or be produced as a result of dissociative adsorption of hydrogen molecule at the epoxy site. The DFT calculations showed that the curvature of the CNT would affect the rates of the reaction. Experimental results on SWNT and graphene will also be presented and compared with the DFT findings. Finally the comparison of the experimental data and the DFT predictions on nitrogen doped CNTs as well as the performance of these caatlysts loaded on various substrates will be presented. The data will show that these alternate, renewable catalysts can efficiently effect the hydro-deoxygenation of CO and assist in subsequent chain growth.

(693g) Hierarchical Zeolite Supported Cobalt for the Direct Production of Liquid Fuels From Syngas: Effect of Zeolite Structure

Sartipi, S., Delft University of Technology
Kapteijn, F., Delft University of Technology
Gascon, J., Delft University of Technology
Originally presented on: 11/7/2013 14:25:00 - 14:43:00

Hierarchical zeolite supported cobalt for the direct production of liquid fuels from syngas: effect of zeolite structure

S. Sartipi,* Freek Kapteijn and Jorge Gascon

Catalysis Engineering, Department of Chemical Engineering, Delft University of Technology, Julianalaan 136, 2628 BL Delft, The Netherlands. Fax: +31 15 2785006; Tel: +31 15 2786733; E-mail: S.Sartipi@tudelft.nl

Keywords: Catalysis, Fischer-Tropsch; hydrocracking; mesoporous zeolite

Introduction

Fischer-Tropsch Synthesis (FTS) is the core of the Gas-to-Liquid (GTL) technology. During FTS, a mixture of CO and H_2 (syngas) is catalytically converted into a wide spectrum of hydrocarbon chains through a surface polymerization reaction. GTL technologies are currently economically attractive only at very large scales. In this sense, process intensification is a must if these tools are to be applied for applications where limited feedstock availability (*e.g.*, Biomass-to-Liquid (BTL)) or spatial constraints (*e.g.*, conversion of offshore flare gas) are obstacles [1]. GTL process intensification has been investigated in terms of both reactor and catalysis engineering: while the use of structured catalysts and reactors maximizes the FTS efficiency by improving mass and energy transport [2–5], many efforts have been devoted to developing catalyst formulations able to maximize the direct production of liquid fuels (desired products) by combination of FTS and hydrocarbon cracking and isomerization into one single catalyst particle [6]. A promising approach consists of using mesoporous (hierarchical) zeolites as FTS catalyst supports [6–8]. Bringing zeolite acid sites in a close proximity to the FTS active phase results in bifunctional catalysts exhibiting a high selectivity to gasoline fraction [6]. With most studies dealing with the use of ZSM-5 as catalysts carrier, it has been proposed that hydrocarbon cracking and isomerization contribute to a non-ASF (Anderson-Schulz-Flory) product distribution. However, very little is known about the quantitative effect of both acid catalyzed reactions and even less about the effect of zeolite properties and structure on the product distribution.

In this contribution, different zeolite topologies are modified with respect to their mesoporous hierarchy and amount of extra framework aluminum (EFAl) and quantitatively compared for application in Co-based FTS catalysts. By various spectroscopy techniques and acid catalyzed hydrocarbon conversion reactions, it is demonstrated that the zeolite topology and, most importantly, the amount and strength of acid sites are key parameters in the performance of bifunctional catalysts to maximize the FTS product selectivity towards liquid hydrocarbons.

Experimental

Three selected zeolites in proton form with different framework topologies, namely MFI (H-ZSM-5), delaminated MWW (H-ITQ-2) and FAU (H-USY) were used as FTS catalyst support. H-ITQ-2 was directly used (as support), while commercial H-ZSM-5 and H-USY zeolites were leached in basic solutions to increase their mesoporous surface area. The mesoporous H-ZSM-5 ('*meso*H-ZSM-5') was also acid treated for EFAl adjustments. The (commercial) parent zeolites as well as the modified ones were loaded with 10 to 20 wt.% Co. For comparison, Co/SiO_2 was chosen as a conventional reference catalyst.

The textural and chemical properties of supports and catalysts were thoroughly characterized by various techniques including N_2 physisorption, XRD, ICP-OES and quasi *in-situ* TEM. The effect of base/acid treatments on the zeolites acidity was assessed by NH_3-TPD and FTIR assisted pyridine adsorption. Employing CO as a probe molecule, the nature of metallic surface sites and different Co species in the catalysts were studied by DRIFTS. Reducibility of the catalysts were evaluated by TPR(H_2) and quasi *in-situ* EXAFS.

FTS experiments were performed in a parallel six-flow lab-scale unit [6]. Acid catalyzed reactions were carried out in a continuous flow fixed-bed reactor, employing *n*-hexane (*n*-C6) as a hydrocarbon model compound.

Results and discussion

Desilication results in pore sizes and volumes similar to the amorphous SiO_2, while acid treatment removes the produced extra framework aluminum, preventing the formation of cobalt aluminates. TEM analyses show that the spatial distribution of Co inside the mesopores of hierarchical zeolite is similar to that in the amorphous SiO_2. Time-on-stream (*TOS*) evolution of cobalt-time-yield during 140 h of reaction demonstrates that stability of the zeolite supported catalysts in terms of CO conversion is comparable to that of the conventional Co/SiO_2. Catalytic activities of Co/*meso*H-USY and Co/*meso*H-ZSM-5 are similar and higher than those of Co/H-ITQ-2 and Co/SiO_2 catalysts. The trend in FTS activities can be attributed to the active phase, *i.e.* smaller Co crystallites over the zeolite supports.

A detailed catalyst performance evaluation (in FTS and *n*-C6 conversion) with a thorough characterization of acidity reveal that: (*i*) *meso*H-USY has a low density of weak acid sites which are not capable of catalyzing the hydrocarbon conversion reactions at low-temperature FTS (LTFT) conditions. Therefore, Co/*meso*H-USY shows only the FTS functionality similar as Co/SiO_2, but due to a lower chain growth probability (α), its product distribution is shifted to lower hydrocarbons. (*ii*) H-ITQ-2 has a high density of acid sites active for isomerization. Thus, Co/H-ITQ-2 combines both FTS and acid functionalities and indeed *iso*-to-*normal* C4 ratio over this catalyst is higher than that over Co/*meso*H-USY and Co/SiO_2. Branched hydrocarbons may not participate in the chain propagation as effectively as the linear ones, further decreasing the chain growth probability. Although wax production can be eliminated at a lower α, this will lead to increased selectivities towards C1 and C2–C4, which are less desired. (*iii*) *meso*H-ZSM-5 exhibits a stronger acidity (relative to *meso*H-USY and H-ITQ-2) resulting in activity for both isomerization and hydrocracking at LTFT conditions. In this way, hydrocracking of primary FTS products takes place on Co/*meso*H-ZSM-5. As a result, the produced fraction of gasoline range hydrocarbons is considerably larger over Co/*meso*H-ZSM-5 than that over the other catalysts.

Among *meso*H-USY, H-ITQ-2 and *meso*H-ZSM-5 supported catalysts, only Co/*meso*H-ZSM-5 displays a significant deviation from an ASF product distribution. Hydrocarbon isomerization alone cannot explain a non-ASF distribution, pointing to the crucial role of cracking activity in the performance of bifunctional FTS catalysts.

As compared with the conventional catalyst Co/SiO_2, Co/*meso*H-ZSM-5 is three times more selective to C5–C11 after 140 h *TOS* and two times more active, while producing far less wax (C21+). On the other hand, the strong Co-zeolite interaction as revealed by TPR(H_2) and quasi *in-situ* EXAFS results in the stabilization of lower coordinated Co sites (which are characterized by IR assisted CO adsorption) and in a higher selectivity towards methane. *n*-C6 conversion reactions suggest that the latter is due to increased activity for hydrogenation and hydrocarbon hydrogenolysis reactions at such coordinatively unsaturated Co sites.

Conclusions

Co catalysts supported on hierarchical zeolites show a high and stable catalytic activity in Fischer-Tropsch Synthesis during 140 h time-on-stream. Zeolite acid density and strength are essential parameters to tune the FTS product selectivity towards liquid hydrocarbons. Only strong acid sites, active for hydrocracking at the operating temperature window of Co-based FTS catalysts, give rise to deviations from a conventional ASF product distribution. These results highlight the importance of the zeolite structure in the development of intensified processes for the direct production of liquid fuels from syngas.

References

[1] C. Kibby, K. Jothimurugesan, T. Das, H.S. Lacheen, T. Rea, R.J. Saxton, Catal. Today, In Press.

[2] R.M. de Deugd, F. Kapteijn, J.A. Moulijn, Top. Catal. 26 (2003) 29–39.

[3] F. Kapteijn, R.M. de Deugd, J.A. Moulijn, Catal. Today 105 (2005) 350–356.

[4] T.J. Schildhauer, K.V. Pangarkar, J.R. van Ommen, J. Nijenhuis, J.A. Moulijn, F. Kapteijn, Chem. Eng. J. 185–186 (2012) 250–266.

[5] D. Vervloet, F. Kapteijn, J. Nijenhuis, J.R. van Ommen, Catal. Sci. Technol. 2 (2012) 1221–1233.

[6] S. Sartipi, J. van Dijk, J. Gascon, F. Kapteijn, Appl. Catal. A-Gen. 456 (2013) 11–22.

[7] S. Sartipi, K. Parashar, M. Makkee, J. Gascon, F. Kapteijn, Catal. Sci. Technol. 3 (2013) 572–575.

[8] S. Sartipi, M. Alberts, M. J. Meijerink, T.C. Keller, J. PÃ©rez-Ramírez, J. Gascon, F. Kapteijn, ChemSusChem, In Press, doi:10.1002/cssc.201300339.

(693h) High Fischer-Tropsch Performance of Cobalt Catalyst Supported On Nitrogen-Doped Bamboo-Shaped Carbon Nanotubes

Fu, T., Tianjin University
Li, Z., Tianjin University
Originally presented on: 11/7/2013 14:43:00 - 15:01:00

High Fischer-Tropsch Performance of Cobalt Catalyst Supported on Nitrogen-Doped Bamboo-Shaped Carbon Nanotubes

Tingjun Fu, Zhenhua Li

Key Laboratory for Green Chemical Technology of Ministry of Education, School of Chemical Engineering & Technology, Tianjin University, Tianjin, China

The recent renewed interest in Fischer–Tropsch (FT) synthesis has largely been driven by the growing demand for clean fuels and the utilization of remote sources of natural and associated gas. Supported cobalt catalysts have been considered the preferred catalyst for the conversion of synthesis gas to long-chain paraffin. Typical FT synthesis catalyst supports are oxide supports like Al2O3, SiO2 and TiO2. However, the use of these supports can generate a strong metal-support interaction and lead to the formation of hardly reducible compounds. As an inert material, carbon materials display special properties like high purity, high mechanical strength, good electrical conductivity, high thermal stability and large surface area. In contrast to oxide supports, only weak interaction between the metal and carbon is evident on carbon supported cobalt catalysts, which can improve FT performance and allow better study of the intrinsic properties of the cobalt active sites.

In this work, bamboo-shaped multiwalled carbon nanotubes doped with nitrogen (NB-CNTs) was used as support for preparing the cobalt catalyst. The activity studies revealed that the acid treatment can significantly enhance the activity of the cobalt catalyst supported on NB-CNTs. Nitrogen incorporation and the surface groups formed during the acid treatment have deep influence upon carbon structure, cobalt dispersion and reducibility of the cobalt catalyst. Compared with cobalt catalysts supported on CNTs, NB-CNTs supported cobalt catalyst has an increased activity for FT synthesis though it has smaller surface area.

Keywords: Fischer-Tropsch synthesis, Cobalt catalyst, Nitrogen doped bamboo-shaped multiwalled carbon nanotubes

(760a) Response of Low-Permeability Carbonate Samples to CO2-Brine Exposure: Experiments & Modeling

Smith, M., Lawrence Livermore National Laboratory
Hao, Y., Lawrence Livermore National Laboratory
Carroll, S. A., Lawrence Livermore National Laboratory
Originally presented on: 11/7/2013 15:15:00 - 15:40:00

Several large-scale subsurface carbon utilization and sequestration projects have been implemented in response to rising levels of atmospheric carbon dioxide (CO_2). Good candidate scenarios include former CO_2-enhanced-oil-recovery fields, which can be transitioned into long-term CO_2 storage sites given the presence of a sufficiently impermeable "seal" formation. Because of the high reactivity of carbonate minerals such as calcite and dolomite under CO_2-acidified conditions, carbonate reservoir formations can be expected to respond to CO_2 injection near the injection zone with generation of new pore space (and potential permeability enhancement), as CO_2-charged fluids partially dissolve carbonate host minerals until chemical equilibrium is reached. However, this same type of reaction can also lead to the formation of less desirable fast fluid pathways if the seal formation also includes carbonate minerals, and these preferential pathways could allow leakage of CO_2 into overlying geologic units. A full understanding of the impact of chemical and mechanical stresses (e.g., fluid-mineral disequilibria, fluctuating reservoir conditions, etc.) resulting from CO_2 injection, and description and/or prediction of how these stresses will translate into meaningful changes in permeability throughout the storage reservoir, is an ongoing active area of research.

To address these questions, we present results from several core-flood experiments, conducted at *in situ* storage temperature and pressure, on well-characterized low-permeability carbonate core samples from two different carbon storage sites (Weyburn, Saskatchewan, Canada; Wellington, Kansas, USA). The purpose of these experiments was to explore the effects of physical (pore versus mineral distribution, connectivity of pore space) and chemical (variable $CO_{2(aq)}$ concentrations, carbonate versus non-carbonate mineral distribution) heterogeneity on the evolution of permeability within these core samples. To quantify these attributes, each sample was imaged via X-ray microtomography both prior to and after each brine-CO_2 core-flooding experiment. Geochemical fluid samples were also collected throughout each experiment and interpreted to define each sample's approach to specific fluid-mineral saturation states. The range of initial bulk permeability of these carbonate samples spans at least five orders of magnitude; however, it is evident that it is not necessarily the initial permeability value but rather the contrast in permeability between neighboring regions that controls the ultimate evolution of fluid flow within these samples. In addition to the characterization and experimental data, we also present reactive transport model simulations, which were conducted to calibrate the relationships between porosity, permeability, and carbonate dissolution rate laws. Our results, to date, indicate that a continuum-scale model can be used to describe the evolution of permeability in a reactive carbonate-brine-CO_2 system, provided that effective permeability can be reproduced well at the model element scale. Correlations between changing porosity and permeability values can also be constrained using the available experimental data. However, we find that even at the high resolution of the tomography data, difficulties arise in adequately describing the effective reactive specific surface area of minerals within these samples. Thus, additional adjustment of kinetic rate constant values is necessary in order to account for order-of-magnitude uncertainties in surface area estimations.

(760b) CO2 Leaks From Geological Storage: Geomechanics, Fluid Flow and Phase Transitions

Gor, G., Rutgers, The State University of New Jersey
Prevost, J. H., Princeton University
Originally presented on: 11/7/2013 15:40:00 - 16:05:00

Deep saline aquifers are considered as a promising option for long-term storage of carbon dioxide. However, risk of CO_2 leakage from the aquifers through faults, natural or induced fractures or abandoned wells cannot be disregarded. Therefore, modeling of various leakage scenarios is crucial when selecting a site for CO_2 sequestration and choosing proper operational conditions.

Carbon dioxide is injected into wells at supercritical conditions ($t^\circ > 31.04$ °C, $P > 73.82$ bar), and these conditions are maintained in the deep aquifers (at 1-2 km depth) due to hydrostatic pressure and geothermal gradient. However, if CO_2 and brine start to migrate from the aquifer upward, both pressure and temperature will decrease, and at the depth of 500-750 m, the conditions for CO_2 will become subcritical. At subcritical conditions, CO_2 starts boiling and the character of the flow changes dramatically due to appearance of the third (vapor) phase and latent heat effects.

When modeling CO_2 leaks, one needs to couple the multiphase flow in porous media with geomechanics. These capabilities are provided by Dynaflow, a finite element analysis program [1]; Dynaflow has already showed to be efficient for modeling caprock failure causing CO_2 leaks [2, 3]. Currently we have extended the capabilities of Dynaflow with the phase transition module, based on two-phase and three-phase isenthalpic flash calculations [4]. We have also developed and implemented an efficient method for solving heat and mass transport with the phase transition using our flash module. Therefore, we have developed a robust tool for modeling CO_2 leaks.

In the talk we will give a brief overview of our method and illustrate it with the results of simulations for characteristic test cases.

References:

[1] J.H. Prevost, DYNAFLOW: A Nonlinear Transient Finite Element Analysis Program. Department of Civil and Environmental Engineering, Princeton University, Princeton, NJ. http://www.princeton.edu/~dynaflow/ (last update 2012), 1981.
[2] M. Preisig, J.H. Prevost, Coupled multi-phase thermo-poromechanical effects. Case study: CO2 injection at In Salah, Algeria, International Journal of Greenhouse Gas Control, 5 (2011) 1055-1064.
[3] G.Y. Gor, T.R. Elliot, J.H. Prevost, Effects of thermal stresses on caprock integrity during CO2 storage, International Journal of Greenhouse Gas Control, 12 (2013) 300-309.
[4] M.L. Michelsen, J.M. Mollerup, Thermodynamic Models: Fundamentals and Computational Aspects. 2nd Edition, Tie-Line Publications, 2007.

(760c) Modeling of Wellbore Cement Core-Flood Experiments

Myint, P. C., Lawrence Livermore National Laboratory
Walsh, S. D. C., Lawrence Livermore National Laboratory
Du Frane, W. L., Lawrence Livermore National Laboratory
Mason, H. E., Lawrence Livermore National Laboratory
Hao, Y., Lawrence Livermore National Laboratory
Carroll, S. A., Lawrence Livermore National Laboratory
Originally presented on: 11/7/2013 16:05:00 - 16:30:00

We present a theoretical model of our core-flood experiments in which brine saturated with carbon dioxide (CO_2) is injected through fractured core samples composed of wellbore cement mated with caprock. When exposed to the CO_2-saturated brine, the cement undergoes a series of reactions that leads to the dissolution of minerals like calcium into the brine. These reactions have implications for the long-term efficacy of subsurface CO_2 sequestration, because they may affect the fracture permeability and the mechanical properties of the wellbore cement. The effluent chemistry data from the experiments show that the excess calcium concentration obey a 'super-diffusive' growth. We find that a good match with experimental data can be achieved by introducing a term that represents a constant background source of calcium. Our model provides explanations for the physical origin of this source term.

(760d) Evaluation and Characterization of Gas Shale Pore Space Using Low-Pressure Gas Adsorption for CO2 Sequestration Applications

Rupp, E. C., National Energy Technology Laboratory
Jew, A., Stanford University
Lee, K., Stanford University
Wilcox, J., Stanford University
Originally presented on: 11/7/2013 16:30:00 - 16:55:00

The need for practical solutions to both carbon capture and carbon sequestration becomes imperative as the climatic impact of anthropogenic CO_2 emissions become increasingly clear. With current global CO_2 emissions increasing beyond 30 gigatons per year, all potential CO_2 reservoirs should be evaluated for sequestration applications. The current focus on non-conventional natural gas resources has the potential to free intriguing reservoirs for sequestration. Many questions, including CO_2 capacity and permeability, exist regarding the viability of shale reservoirs for CO_2 storage and studying the pore space due to the heterogeneity of shale. In addition, a better understanding of organic-rich shale can increase the understanding of non-hydrocarbon producing shale, which are considered to be important cap rocks for traditional reservoirs.

Low-pressure gas adsorption has been used to evaluate shale samples from active gas plays, including the Haynesville and Eagle Ford formations. Adsorption and desorption isotherms were created for the entire sample using N_2 at 77 K, Ar at 87 K and CO_2 at 273 K and evaluated for both total pore volume and pore size distribution using density functional theory models. The samples were then partitioned into 4 size bins of 100 μm each, and additional isotherms were performed. In this manner, changes in pore volume and pore size distribution can be deduced. Further analysis was performed to support the gas adsorption observations, including Fourier transform infrared spectroscopy (FTIR), scanning electron microscopy (SEM), elemental analysis and total and inorganic carbon contents.

(760e) Geomechanical and Geochemical Influences On the Permeability of Wellbore Cement Fractures Exposed to CO2-Rich Brine: An Experimental and Modeling Study

Carroll, S., Lawrence Livermore National Laboratory
Walsh, S., Lawrence Livermore National Laboratory
Mason, H., Lawrence Livermore National Laboratory
Du Frane, W., Lawrence Livermore National Laboratory
Originally presented on: 11/7/2013 16:55:00 - 17:20:00

The objective of this work was to explore and quantify the relationships between chemical alteration, deformation, and permeability at the wellbore/caprock interface important to long-term geologic carbon storage through experiment and modeling. The core flood experiments span variable pCO_2, flow rate, and cement – caprock fracture apertures at 60°C and 24.8 MPa. We used x-ray computed micro tomography to spatially resolve the fracture surface and the extent of alteration, three dimensional digital image correlation to spatially resolve plastic deformation resulting from chemical alteration, and time dependent solution chemistry and pressure to track coupled evolution of chemical alteration and mechanical deformation. The experimental data are used to constrain models of wellbore/caprock interface using a numerical simulator that couples reaction front chemistry, flow and transport of dissolved species, and elastic and plastic deformation.

Coupled processes between chemical alteration and material compressibility were largely responsible for decrease in permeability despite measured porosity increase at the cement/caprock interface as a result of the chemical alteration. Our results show that CO_2-rich brines alter wellbore cement into distinct portlandite depleted, carbonate, and aluminum-bearing amorphous silicate layers and change the compressibility of cement at the wellbore/caprock interface. Furthermore, dissolution of portlandite from the cement resulted in plastic deformation of the surface contacts enabling fluid flow paths. The deformation restricts flow paths and lowers the overall permeability. Chemical alteration of the cement was controlled by diffusion of reactants and products through cement and its alteration layers, as well as calcite equilibrium at the carbonate/amorphous silicate boundary and analcime equilibrium at cement/caprock interface.

(760f) Forsterite Dissolution From Indian Olivine in Saline Water

Agrawal, A. K.,
Giammar, D. E.,
Mehra, A., Indian Institute of Technology, Bombay (IITB)
Originally presented on: 11/7/2013 17:20:00 - 17:45:00

The rate and mechanism of forsterite dissolution was studied using naturally weathered olivine FO_{79} $(Mg_{1.591}Fe_{0.182})SiO_4$ from an Indian source, that also contains serpentine mineral lizardite. A series of batch dissolution experiments were carried out to check the influence of temperature (25– 75 °C), initial olivine concentration (0.5 and 20 g/L) and salinity (0– 35,000 mg/L NaCl) under fixed headspace CO_2 pressure (P_{CO2}= 1 bar) on forsterite dissolution. Dissolved Mg, Si and Fe concentrations were determined by inductively coupled plasma atomic emission spectroscopy. End-product solids were characterized by scanning electron microscopy and X-ray diffraction. Initially, rates of dissolution of Si and Mg were observed to be in stoichiometric proportion. After 8 h the dissolution rate was observed to decline. At the end of the experiment (504 h) an amorphous silica-rich layer was observed on the forsterite surface. This results in decay of the dissolution rate. A comparison was made between experimentally obtained dissolution rates and corresponding theoretical estimates from a batch dissolution model.

(787a) 4CU: A Comprehensive, Coordinated Approach to Carbon Capture and Utilisation â€" Electrocatalytic Co-Reduction

Elder, R. H., University of Sheffield
Allen, R. W. K., University of Sheffield
Originally presented on: 11/8/2013 8:30:00 - 8:49:00

Converting carbon dioxide to vehicle fuel has the potential to reduce future dependence on fossil fuels. In effect a carbon neutral fuel cycle is created: carbon dioxide is captured and converted to fuel then that fuel is burned to form carbon dioxide. The challenge is in finding processes capable of converting the carbon dioxide efficiently using renewable energy sources.

4CU is a 4.5 year, £5.7m collaborative project between the University of Sheffield, Queens University Belfast, University College London and Manchester University. It is the first project of its kind and aims to investigate carbon capture and utilisation at a systems level, including comprehensive life cycle and process analyses; at a detailed technology level, including several process routes; and at a micro level, using quantum mechanical and classical computational modelling techniques. As an example system we have chosen the anaerobic digestion of wet waste, followed by CO_2 capture using both novel processes and chemicals: microbubble systems and ionic liquids, including those that are liquid, supported and solid. The CH_4 stream will be burned in a conventional CHP, the CO_2 stream from which will be captured using similar methods to above. Conversion of the captured CO_2 is then investigated using several different technologies: electrocatalysis, plasmolysis and catalysis. An overview of the project, including preliminary results, will be presented.

High temperature co-electrolysis of steam and carbon dioxide using a solid oxide cell (SOC) has been shown to effectively produce syngas (CO and H_2), which can then be converted to fuel using Fischer Tropsch technology. 4CU Sub Project 5 concentrates on fundamental understanding and materials advances for optimising the co-electrolysis process. Results from in situ spectroscopy, including Raman and DRIFTS, as well as thermal imaging and impedance characterisation will be presented.

(787b) Carbonate Ceramics and Concrete: A Disruptive Technology for CO2 Utilization and the Construction Business

Riman, R. E., Rutgers, The State University of New Jersey
DeCristofaro, N., Solidia Technologies
Blinn, K.,
Originally presented on: 11/8/2013 8:49:00 - 9:08:00

Alternatives to Carbon Capture and Storage (CCS) are needed that can use captured CO_2 directly or convert it to valuable products, such as fuels, chemicals, plastics, or building materials. This paper will describe a breakthrough process invented and patented by Rutgers called hydrothermal liquid phase sintering. When CO_2 is incorporated into the hydrothermal liquid phase to react with ultramafic mineral silicates at temperatures below 100˚C, chemically bonded ceramics can be produced with outstanding properties. Our work has demonstrated that both building and infrastructure materials can be manufactured that meet or exceed property specifications for a range of applications to justify commercialization. At the same this process is green in terms of its carbon footprint and energy consumption. CO_2 mass balance computations suggest that use of mineral-based $CaSiO_3$ can yield a truly carbon-negative material that consumes a fraction of the energy required for cement or concrete processing. Solidia Technologies was founded as a Rutgers start-up to build construction materials businesses that yield materials with exceptional performance and cost while maintaining its green profile. Progress towards commercialization and its potential impact on global CO_2 emissions and energy will be discussed. This presentation will also provide a perspective with respect to the sustainability of this process relative to stone, ceramic and concrete processing.

(787c) Maximizing CO2 Uptake in Concrete Blocks Carbonation
Shao, Y., McGill University
Originally presented on: 11/8/2013 9:08:00 - 9:27:00

Maximizing CO_2 uptake in concrete blocks carbonation

Yixin Shao[1]*, Yaodong Jia[1], Zhen He[2] and Liang Hu[3]

[1]Department of Civil Engineering, McGill University, Montreal, Canada H3A 2K6

[2]School of Water Resource and Hydropower, Wuhan University, Wuhan, China 430072

[3]3H Company, Lexington, KY, USA 40511

*Corresponding author: yixin.shao@mcgill.ca

Abstract:

Portland cement based concrete products are well known by its reactivity with carbon dioxide. It is the calcium compounds in cement that react with CO_2 and produce thermodynamically stable calcium carbonates. It was estimated at full carbonation when all CaO reacts with carbon dioxide, CO_2 uptake by cement could reach to 50% by cement mass, i.e. one ton of cement could uptake half a ton of CO_2. Full carbonation could happen in weathering carbonation of concrete in more than 30 years when atmospheric CO_2 slowly but progressively reacts with hydration products in matured concrete. Nevertheless it is not a desired reaction since hydration products are decomposed and strength of concrete is reduced. On the other hand, early carbonation during concrete curing provides an ideal process window for CO_2 utilization. It is an accelerated hydration and can significantly improve durability performance of concrete. Unlike steam curing which is well commercialized, early carbonation curing has never been used in large scale. One of the limiting factors is the cost of CO_2 gas. In near future, high purity low cost CO_2 could be available in large quantity as the result of emission regulation. This provides an opportunity for commercial application of carbon dioxide in early concrete carbonation which will integrate carbon capture, utilization and storage (CCUS) into one process and will also show economical and technical benefits.

This paper is to present a new carbonation process that incorporates a hydration presetting prior to early carbonation. Concrete products with high water to cement ratio and formed by vibration compaction can be used as candidate products for carbon capture and utilization. Since CO_2 uptake capacity is unique for each product and its associated process, concrete block will be used as an example to prove the concept. Concrete blocks have been widely utilized in construction. The US market for concrete blocks and bricks is projected to increase to 4.3 billion units per year in 2014. Concrete blocks are made of calcium silicate cement, porous in nature and mass produced. The block plants are also demographically distributed and strategically located close to CO_2 sources such as power plants and cement plants. Currently concrete blocks are produced by steam curing. If steam process can be replaced by CO_2 curing, concrete blocks can serve as CO_2 sink for carbon storage through utilization. It is also impetus to show energy comparison between steam and CO_2 systems.

The paper will first introduce the concrete block carbonation process that involves hydration and carbonation in different sequence, based on which a reaction model is proposed to estimate theoretical maximum CO_2 uptake within the process window. The reaction kinetics of four cement phases: tricalcium silicate ($3CaO{\cdot}SiO_2$ or C_3S), dicalcium silicate ($2CaO{\cdot}SiO_2$ or C_2S), tricalcium aluminate ($3CaO{\cdot}Al_2O_3$ or C_3A), tetra calcium aluminoferrite ($4CaOAl_2O_3Fe_2O_3$ or C_4AF), and four hydration products: calcium hydroxide, calcium-silicate-hydrate, calcium-aluminate-hydrate, calcium aluminate ferrite hydrate, are studied. Based on the reaction kinetics, a model is further verified by cement solution carbonation which represents the ideal reaction conditions and also by normal-weight and lightweight concrete carbonation.

Theoretical maximum of CO_2 uptake within a 24-hour process window was estimated and compared with experimental CO_2 uptake by concrete blocks carbonation. The combination of hydration presetting and carbonation helped enhance reaction efficiency. Hydration presetting of 4-18h could remove 40-60% of free water to create path for CO_2 diffusion, make space for carbonate precipitation and produce early hydration products for carbonation. The best CO_2 uptake was obtained with relatively high water content in initial mix design followed by 18h presetting and 2 - 4h carbonation. It was 18.3% in normal weight concretes and 24.2% in lightweight concretes. This process window is practically feasible.

Theoretical maximum of CO_2 uptake was estimated based on assumption that at early carbonation, only C_3S, C_2S and their hydration products were reactive with CO_2. The contribution by aluminate and ferrite phases was neglected. It was also assumed that after hydration presetting, the remaining C_3S and C_2S were fully carbonated into C-S-H and $CaCO_3$. Based on these assumptions, theoretical maximum CO_2 uptake was computed. It was 19.7% with no presetting and 27.1% with 18h presetting. The results were in agreement with cement solution carbonation. CO_2 uptake in normal weight concrete reached 67% reaction efficiency after 2h carbonation and CO_2 in lightweight concrete reached 89% after 4h carbonation. The higher uptake rate in lightweight concrete was also attributed to reactive slag aggregates. If uptake by aggregates was subtracted from total uptake, lightweight concrete had shown a close value to normal weight concrete. Carbonation at prolonged duration and at high pressure could promote more CO_2 uptake. In these cases, C-S-H created by carbonation as well as calcium alunimate and calcium aluminate ferrite phases could be carbonated too. However, full carbonation with 50% CO_2 uptake was never reached even after continued reaction for two days.

Concrete blocks carbonation will promote the value of CO_2 as a commodity for building industry and will permanently store CO_2 in thermodynamically stable calcium carbonates. Concrete block is only one of numerous cement-based building products. The annual production of Portland cement in US is about 100 mt. If half of its products could be treated by carbonation at a CO_2 uptake rate of 10-25%, CO_2 utilization by concrete industry alone could reach 5-12 mt/year.

(787d) Dehydrogenation of Ethylbenzene With CO2 Over Fe2O3/Al2O3-ZrO2 Catalyst: Influence of Fe2O3 Content

Zhang, S., Taiyuan University of Technology
Li, W., Taiyuan University of Technology
Originally presented on: 11/8/2013 9:27:00 - 9:46:00

Recently, styrene has been produced by ethylbenzene dehydrogenation over iron oxide catalyst with a large excess of superheated steam. In order to improve ethylbenzene conversion and enhance thermal efficiency, ethylbenzene dehydrogenation to produce styrene with CO_2 (EDSC) is developed. Compared to use H_2O, using CO_2 is an obvious energy-saving process and conformed to the trend of green chemistry. However, the present commercial catalyst is ineffective in the presence of CO_2. The indispensable requirement is to select a proper catalyst system for the EDSC process.

In our early research, we have found that Al_2O_3-ZrO_2 (AZ) support is proper for the EDSC catalyst system, and its best performance is ZrO_2 at 10 wt% (AZ10). For a catalyst, active component content affects its structure and then influences its catalytic activity. Therefore, finding the optimal addition mount of active component, then exploring catalyst characteristic performance and analyzing the reaction mechanism with this catalyst, which is meaningful for a catalyst system. Therefore, using AZ10 as the support, we prepare Fe_2O_3/Al_2O_3-ZrO_2 (FAZ) catalyst, select Fe content as x (2-10) mmol in 1g AZ support and the resultant catalyst is demoted as FxAZ10.

Based on the FAZ catalyst activity evaluation result, F4AZ10 catalyst has acquired the best catalytic activity with 61.2% ethylbenzene conversion and 90.4% styrene selectivity at 550 °C. Through XRD and TEM-EDS analysis, fresh FAZ catalysts are in X-ray amorphous state or displayed only weak and broad lines of alumina, but used FAZ catalysts appear Fe_3O_4 (220) crystalline phase. In addition, from F4AZ10 to F10AZ10, the higher of the Fe_2O_3 content in FAZ catalyst, the more obvious of Fe_3O_4 crystalline phase after the reaction. Part of active sites reduction, agglomeration and catalyst structure change seem to cause the catalyst deactivation. N_2 physical adsorption displays that both S_{BET} and pore volume are decreased after being used for 8 hours. Raman spectroscopy indicates the carbon characteristic peaks at used F4AZ10 catalyst. These characterization results could be concluded that carbon formation and deposition is also an important factor of catalyst activity decline. According to the above analysis, the reaction mechanism of FAZ catalyst is also represented.

(787e) Feasibility Analysis of the Implementation of a CO2 to Synthetic Fuel Process in Offshore Oil Production Platform

Santos, R. A. D., Petroleo Brasileiro SA - PETROBRAS
Le Roux, G. A. C., Polytechnic School of the University of SÃ£o Paulo
Originally presented on: 11/8/2013 9:46:00 - 10:05:00

The discovery of a new oil province known as pre-salt, located in the Brazilian coast, represents a new frontier in the world's oil production.

Amongst other challenges involved with the exploration and production in that region, the CO_2 concentration at high levels in the fluids of some of those reservoirs has been the most relevant challenge for the companies that explore and produce in that area [2]. This is due to the fact that when the CO_2 is found in natural gas at more than a substantial proportion, it may corrode the transport and transfer ducts, along with atmospheric pollution related to the greenhouse effect and gas commerce barriers for not satisfying the requirements of the National Petroleum Agency (ANP), etc. [1]

Once the natural gas is extracted, the amount of CO_2 produced can be both injected or stored in depleted reservoirs, saline aquifers and carbon reservoirs or it can be reinjected in production fields of secondary oil recovery [1], being this the model adopted by the platforms responsible for the pre-salt production today.

An alternative to sequestering the CO_2 extracted from natural gas is using it as a supply of raw material or as an input to the synthesis of chemical products, especially those with great market demand. From that perspective, the research and the development of processes and technologies that aim the chemical conversion of the CO_2produced in oil platforms into products with a higher aggregate value presents itself as an excellent alternative to guarantee the environmental preservation, associated with its profitability obtained from what is considered to be the main cause of global warming.

One of the products that can be obtained indirectly from CO_2 is synthetic fuel, produced by the process known as *Gas to Liquids*, or GTL. In the GTL process, the synthesis gas (H_2 and CO) is firstly produced by the reaction of natural gas (CH_4) with CO_2or water, also known as dry reforming and steam reforming, respectively. Next, the synthesis gas is converted into a mixture of hydrocarbons composed essentially of long-straight chain paraffin hydrocarbons and α-olefins.

In this work the technical, economical and potential feasibility of CO_2 capture is analyzed, as well as the production of synthetic fuel using the GTL process, offshore, from carbon dioxide (CO_2) and methane (CH_4), found in the waste chain of the gas treatment in an oil and gas production platform, through dry and steam reforming, followed by Fischer-Tropsch synthesis.

Thus, based on literature data and on the use of a commercial process simulator, the simulation of the suggested process was developed and different alternatives to the reutilization of the waste chains were analyzed, including the recycling and burning of some effluent streams. Techniques of energetic integration, optimization and economic analysis of process have also been applied.

Preliminary results indicate that it is technically and economically possible to produce up to 500 m^3/d of synthetic fuel from effluent CO_2 of primary petroleum processing, using GTL, with CO_2 conversion yields of approximately 57,5% and a carbon efficiency up to 39%.

1. FORMIGLI, JosÃ©. Pre-Salt Reservoirs Offshore Brazil: Perspectives and Challenges. ENERGY CONFERENCE. Miami, 2007

2. IPCC - Intergovernmental Panel on Climate Change, Special Report on Carbon Dioxide Capture and Storage, 2005

(787f) Conversion of Fusel Oil Alcohols to Alkyl Carbonates By Carbon Dioxide Fixation

Pereira, F. S., FCT - UNESP
Pereira, L., FCT - UNESP
Credito, D.,
Girao, L.,
Gonzalez, E.,
Originally presented on: 11/8/2013 10:05:00 - 10:24:00

The development of suitable methods for the preparation of interesting CO_2-containing compounds, like organic carbonates, is an alternative to recycling CO_2 and using it as a substitute for the highly toxic phosgene and its derivatives.[1–4] The selectivity and nature of the capture products are two of the most important aspects of the CO_2 fixation.[5,6] Organic carbonates can be synthesized by reaction of alcohols with CO_2 under catalytical conditions.[7,8] This process is relevant since both CO_2 and fusel oil are by-products from the sugar cane and ethanol production. Alkylcarbonates among other products have wide industrial applications, such as chemical,[9] medicinal[10] and fuel additives.[11] The present work describes the preparation of alkyl carbonates from reaction of fusel oil alcohols with carbon dioxide in the presence of 1.8-diazabicycloundecene (DBU) and 1.5-diazabicyclononene (DBN) and an alkylating agent. The capture of CO_2 by alcohols was carried out with nucleophilic activation of the isopentyl and isobutyl alcohols with the amidines DBU or DBN. The carbon dioxide fixation was accomplished by subsequent alkylation under pressurized-CO_2 conditions. The synthesis of butyl-isopentylcarbonate and butyl-isobtuylcarbonate were confirmed by FT-IR ana, lysis and GC-MS with EI ionization mode. In order to establish the conditions for better yields of the alkylcarbonates, it was investigated the influence of temperature, concentration of reagents, time of reaction and gas pressure in the synthesis.

[1] E. R. Perez, M. O. da Silva, V. C. Costa, U. P. Rodrigues-Filho and D. W. Franco, *Tetrahedron Lett.*, 2002, **43**, 4091–4093.

[2] P. Alessio, D. M. Ferreira, A. E. Job, R. F. Aroca, A. Riul, C. J. L. Constantino and E. R. P. González, *Langmuir*, 2008, **24**, 4729–4737.

[3] C. R. Gomes, D. M. Ferreira, C. J. L. Constantino and E. R. P. Gonzalez, *Tetrahedron Lett.*, 2008, **49**, 6879–6881.

[4] J.M. Hooker,A. T.Reibel, S.M.Hill, M. J. Schueller and J. S. Fowler, *Angew. Chem., Int. Ed.*, 2009, **48**,

3482–3485.

[5] Pereira, F. S.; DeAzevedo, E. R.; Silva, E. F.; Bonagamba, T. J.; Agostini, D.; Magalhães, A.; Job, A. e PÃ©rez, E. R. *Tetrahedron*, 2008, **64**, 10097-10106.

[6] Pereira, F. S.; Agostini, D.; Santo, R.D.E.; deAzevedo, E. R.; Bonagamba, T.J.; Job, A.; PÃ©rez, E. R. *Green Chemistry*, 2011, **13**, 2146-2153.

[7] Sankar, M.; Madhavan, C. N.; Murty, K. V. G. K.; Manikandan P. *Applied Catalysis A: General*, 2006, **312**, 108–114.

[8] Park, D. W.; Jeong, E. S.; Kim, K. H.; Bineesh, K. V.; Lee, J. W.; Park, S. W. *Studies in Surface Science and Catalysis,* 2006, **159**, 329–332.

[9] a. Schaffner, B.; Andrushko, V.; Bayardon, J.; Holz, J.; Borner, A. *Chirality*, 2009, **21**, 857-861. b. Vollmer, C.; Thomann, R.; Janiak, C. *Dalton Trans*., 2012, **41**, 9722-9727.

[10] Ghosh, A. K.; Duong, T. T.; Mckee, S. P.; Thompson, W. J. *Tetrahedron Lett*. 1992, **32**, 2781.

[11] Carbon Dioxide Recovery and Utilization, Aresta, M. Ed., Kluwer Academic Publishers, London, 2003.

(787g) CO2 Emission Reduction By the Reverse Boudard Reaction - Coal Gasification With CO2

Feng, M., Southwest Research Institute

Originally presented on: 11/8/2013 10:24:00 - 10:43:00

CO_2 capture and sequestration from coal-fired power plants is an energy intensive process, both in CAPEX and OPEX. CO_2 emission reduction by gasification of coal with CO_2 is an alternative method for future new coal-fired power plants. The reaction of carbon with CO_2 giving CO is also called the reverse Boudouard reaction. This reaction is highly endothermic; however, energy from other co-existing exothermic reactions can be used to reduce the overall energy consumption. We developed a new process for this application [1]. Initial coal gasification under thermal plasma conditions makes the reaction more favorable. With the presence of water vapor, this process is able to produce syngas with $CO_2$and coal as the feed. This paper will also address the subsequent syngas cleanup step.

Reference

[1] Maoqi Feng, Francis Huang, Richard Johnson, Enhancement of syngas production in coal gasification with CO_2 conversion under plasma conditions, US Patent 8,435,478 (issue May 7, 2013).

(787h) Conversion of Carbon Dioxide to Various Carboxylic Acids Or Carbonates Under Mild Conditions

Zhang, Y., IBN

Originally presented on: 11/8/2013 10:43:00 - 11:02:00

Conversion of Carbon Dioxide to Various Carboxylic Acids or Carbonates under Mild Conditions

Yugen Zhang

Institute of Bioengineering and Nanotechnology

31 Biopolis Way, The Nanos, Singapore 138669, Singapore

E-mail: ygzhang@ibn.a-star.edu.sg

Carbon dioxide (CO_2) is a non-toxic, non-combustible, non-flammable gas. The use of carbon dioxide as a renewable and environmentally friendly source of carbon in organic synthesis is a highly attractive approach, but its real world application remains a great challenge. The major obstacles for commercialization of most current protocols are their poor catalytic performance and harsh, severe reaction conditions. It is important to develop new reactions and new protocols for CO_2 transformations at mild conditions and in a cost-efficient manner. This presentation will discuss new approaches for transforming CO_2 to various carboxylic acids on terminal alkynes. Various propiolic acids, substituted acrylic acids and propyl acids have been synthesized with CO_2 in good to excellent yields under very mild conditions. A transition metal-free system for the synthesis five- and six-member ring cyclic carbonate will also be discussed.

American Institute of Chemical Engineers Annual Meeting 2013, Symposium on 'Novel Approaches to CO2 Utilization'.

Direct Synthesis Of Dimethyl Carbonate From Methanol and Carbon Dioxide Using Supported Calcined Hydrotalcite As Catalyst With Ionic Liquid As Promoter Under Supercritical Condition

Kabra, S., Institute of Chemical Technology
Yadav, G.,
Originally presented on: 11/8/2013 11:02:00 - 11:21:00

A sustainable process was developed using supported calcined hydrotalcite catalyst and by avoiding use of harmful organic solvents in the presence of ionic liquid and supercritical CO_2. Present work focuses on direct synthesis of dimethyl carbonate from carbon dioxide and methanol, using supported calcined hydrotalcite as catalyst with phosphonium based ionic liquids as promoter. Different modified Keggin type heteropoly acids supported on hexagonal mesoporous silica were also compared. Ionic liquids are good solvents for many substances and shown good results as promoter for the catalyst. The conversion, selectivity and yields were measured as a function of the amount of reactants, reaction time, temperature, pressure and amount of catalyst. The trihexyl (tetradecyl) phosphonium chloride ionic liquid was found to be good alkali promoter. Lower concentration of methanol, lower temperature and higher pressure favors good conversion and selectivity of desired product DMC. At higher concentrations of methanol there was decrease in conversion as well as selectivity. Higher temperature and lower pressure favors undesired reaction and forms dimethyl ether. The reaction was found to be first order with respect to methanol and the activation energy was calculated as 10.09 Kcal/mol. The catalyst is reusable and can be separated easily. Thus it is a green and environmental benign process.

(816a) Steady-State and Dynamic Modeling of a Moving Bed Adsorber/Regenerator for Solid Sorbent CO2 Capture

Kim, H., National Energy Technology Laboratory
Modekurti, S., West Virginia University
Miller, D. C., U.S. Department Of Energy, National Energy Technology Laboratory
Bhattacharyya, D., West Virginia University
Zitney, S. E., National Energy Technology Laboratory
Originally presented on: 11/8/2013 12:30:00 - 12:50:00

Moving bed reactors hold strong promise for solid sorbent-based CO_2 capture processes by enabling larger working capacities than other reactor types such as fixed- and fluidized-bed reactors. A first-principles model of a moving bed reactor can be used to investigate tradeoffs between various design alternatives and determine the optimal configurations, process conditions, and sorbent types. With this motivation, one-dimensional (1D), two-phase, non-isothermal, steady-state and dynamic models of a moving bed reactor have been developed for use in the U.S. Department of Energy's Carbon Capture Simulation Initiative (CCSI). The models are flexible over a wide range of operating conditions such as gas and solids flowrates and compositions, and operating temperatures. They can be used to represent both an adsorber for CO_2 capture and a regenerator for removing CO_2 from solid sorbents.

In these moving bed reactors, solid sorbent particles move downward due to gravity, contacting a rising gas stream in a counter-current configuration. Since the adsorption reaction is exothermic, an immersed cooler is considered in the adsorber for enhancing solid sorbent loading. An immersed heater is considered in the regenerator to maintain a near-isothermal operating condition. An integral heat-recovery system is used for recovering heat from the hot sorbent leaving the regenerator. The recovered heat is then used to preheat the solid sorbent from the adsorber. The reaction kinetics, heat and mass transfer, and the hydrodynamics are considered in the beds and in the heat recovery system for the regenerator. Sorbent NETL32D, a solid sorbent developed by NETL, is modeled in this work. The kinetic model considers three reactions and has been developed by lumping the internal mass transfer resistances. To capture the mass transfer resistance from/to the bulk to/from the solid sorbent surface, especially in the regenerator that operates at higher temperature, external mass transfer resistances have been modeled. The hydrodynamics of moving beds have not been widely studied, so most coefficients have been derived by analogy with correlations for fixed and fluidized bed systems. The gas flow is modeled by convective gas movement with axial dispersion. For modeling heat exchange between the moving bed and the embedded heat exchangers, heat transfer coefficients have been derived by considering an enhanced packet renewal theory.

The system of partial differential algebraic equations with the appropriate boundary conditions is solved in Aspen Custom Modeler (ACM) using the well-known method of lines. The effect of operating conditions and design parameters on the performance of the adsorber/regenerator reactor has been studied using the steady-state model. Using the dynamic model, a number of disturbances are simulated to study the transient response of the moving bed adsorber/regenerator.

(816b) Modeling, Analysis and Optimization of Process Technologies for Flue Gas Dehydration

Gao, H., Northwestern University
You, F., Northwestern University
Snurr, R., Northwestern University
Originally presented on: 11/8/2013 12:50:00 - 13:10:00

Carbon Capture and Sequestration (CCS) of flue gas from large power plants have received increasing attention because of concerns about the role of CO_2 in global climate change. Among different potential technologies for CCS, adsorption processes based on metal-organic frameworks (MOFs) have received significant attention because MOFs can be designed to achieve very high selectivity of CO_2 over N_2[1]. However, the stability and capacity of MOFs can be seriously affected by the presence of water in the flue gas[2]. Many researchers are working to develop new MOFs that can operate in the presence of water, but here we explore options for removing water from flue gas before a downstream adsorption process. Such a dehydration step can contribute a significant part of the total cost for CO_2 capture[3]. Furthermore, different MOFs may have different tolerance of moisture, so the dehydration requirement (i.e. the water vapor percentage in the dried flue gas), needs to be considered as a variable in order to be connected with possible material selection for the downstream process. To reduce the cost of an adsorption based carbon capture process, the dehydration unit needs careful design and optimization, but this issue is rarely addressed in the existing literature. In this study, we modeled and optimized different process technologies for the dehydration of flue gas. Our goal is to identify the best dehydration technology and minimize the cost of the dehydration process.

We investigated four alternatives for flue gas dehydration, including compression & cooling, cooling & condensation, TEG absorption and solid adsorption. They are compared in terms of cost and separation performance (i.e. loss of other components when satisfying the dehydration requirement). We varied the dehydration requirement from 0.1% to 1% of water in the dry gas for all four technologies. The first three processes are continuous processes and can be well simulated using Aspen Plus. We then apply a response surface method (RSM)[4, 5] to build a function linking the input-output data and minimize the process costs. Adsorption is a dynamic process operated in a cyclic manner[6]. Governing equations form a partial differential equation system, which is fully discretized both in temporal and spatial domain. We apply a first order finite volume method to discretize the spatial domain, and a Radau collocation scheme for temporal discretization. This transfers the original dynamic optimization problem to a nonlinear optimization problem, which can then be modeled and solved in GAMS using general-purpose MINLP solvers to obtain the optimal operating conditions and cost.

Results show the cost for water removal from flue gas can range from a few dollars to tens of dollars per ton CO_2 captured, varying with the dehydration requirement. Among the four technologies, cooling &

condensation proved to be the cheapest process for water removal, but it can only dehydrate to about 0.5% water content with no more than 10% loss of CO_2. TEG absorption loses the least CO_2 but the process cost is higher. The advantages and disadvantages of other technologies are also analyzed and discussed. The consideration of different dehydration requirements provides the foundation for linking with material selection in the downstream adsorption process.

Reference

[1] M. Zhao, A. I. Minett, and A. T. Harris, "A review of techno-economic models for the retrofitting of conventional pulverised-coal power plants for post-combustion capture (PCC) of CO2," *Energy & Environmental Science,* vol. 6, pp. 25-40, Jan 2013.

[2] A. C. Kizzie, A. G. Wong-Foy, and A. J. Matzger, "Effect of Humidity on the Performance of Microporous Coordination Polymers as Adsorbents for CO2 Capture," *Langmuir,* vol. 27, pp. 6368-6373, May 17 2011.

[3] M. M. F. Hasan, R. C. Baliban, J. A. Elia, and C. A. Floudas, "Modeling, Simulation, and Optimization of Postcombustion CO2 Capture for Variable Feed Concentration and Flow Rate. 1. Chemical Absorption and Membrane Processes," *Industrial & Engineering Chemistry Research,* vol. 51, pp. 15642-15664, Dec 5 2012.

[4] J. A. Caballero and I. E. Grossmann, "An algorithm for the use of surrogate models in modular flowsheet optimization," *Aiche Journal,* vol. 54, pp. 2633-2650, Oct 2008.

[5] D. R. Jones, M. Schonlau, and W. J. Welch, "Efficient global optimization of expensive black-box functions," *Journal of Global Optimization,* vol. 13, pp. 455-492, Dec 1998.

[6] R. T. Yang, *Gas separation by adsorption processes*. Boston: Butterworths, 1987.

(816c) On the Design and Modeling of a Rotary Reactor for Chemical-Looping Combustion With a Nickel-Based Oxygen Carrier

Zhao, Z., Massachusetts Institute of Technology
Chen, T., Massachusetts Institute of Technology
Ghoniem, A. F., Massachusetts Institute of Technology
Originally presented on: 11/8/2013 13:10:00 - 13:30:00

Chemical-looping combustion (CLC) is a novel and promising option for carbon capture and sequestration (CCS). Currently almost all the research has been focused on developing CLC-based interconnected fluidized-bed reactors. In this report, a rotary reactor with a nickel-based oxygen carrier (OC) for gas fueled CLC is designed and analyzed. In the reactor, a solid wheel rotates between the fuel and air streams at the reactor inlet and exit. The rotary wheel consists of a large number of micro-channels with nickel oxide coated on the inner surface of the channel walls. The OC oxidizes the fuel in the fuel zone to generate undiluted CO_2 and then it is regenerated in the air zone. Thus pure CO_2 can be readily obtained from the flue streams of the fuel zone and the dilution from the air is avoided. Gas flows through the reactor at elevated pressure and it is heated to high temperature by fuel combustion which can then be used to drive the gas turbine. Design parameters have been selected on the basis of the characteristics of the nickel-based OC and for a thermal capacity of 1MW, the reactor has a diameter of around 1.0m, and a height of about 0.5m. Owing to a relatively slow oxidation rate, an intermediate range of the OC conversion as well as a relatively large share of air sector is used for the design. A one-dimensional plug-flow model has been developed to predict the performances of the reactor. The results indicate that both the fuel conversion and the carbon separation efficiencies are close to unity. In the periodic stationary operation, stable performances are obtained with the gas temperature increasing monotonically from 550℃ to 1200℃. The overall energy in the solid phase is balanced between the reaction heat release, conduction and convective cooling. Sensitivity study has been conducted on the base-case design and the results revealed that the most important parameters are the reduction kinetics, the operating pressure and the feed stream temperatures. Advantages of the rotary design with the nickel-based OC include the compactness, up-scaling feasibility, periodic and stationary operation with limited energy penalty from operation. The potential challenges are also discussed in the report.

(816d) Experimental Characterization of Rotary Reactor for Chemical-Looping Combustion

Chen, T., Massachusetts Institute of Technology
Uddi, M.,
Zhao, Z., Massachusetts Institute of Technology
Ghoniem, A. F., Massachusetts Institute of Technology
Originally presented on: 11/8/2013 13:30:00 - 13:50:00

Chemical looping combustion (CLC) is currently one of the most promising methods to achieve Carbon capture and sequestration (CCS). A novel rotary reactor concept for gas fueled combustion is proposed to eliminate the large energy penalties of traditional CCS technologies during the separation process.[i] The core of the reactor is a solid wheel consisting of numerous micro-channels rotating between fuel and air streams. Metal oxide, known as the oxygen carrier (OC), that reacts with fuel to generate undiluted CO_2 is coated on the inner surface of the channel walls. After the reduction reaction in the fuel zone, the OC is carried over to the air zone by the rotary wheel, where it is regenerated by the oxidation reaction. As a result pure CO_2 can be recycled and sequestrated from the flue streams of the fuel zone by condensing water.

A detailed numerical model has been developed to simulate and validate the capability of the reactor design to completely convert the fuel and capture CO_2 while maintaining desirable stabilities.[i],[ii] However, as the new reactor design employs different reaction mechanism than what is available in literature, experimental investigation is needed. A button cell reactor, which is connected to a mass spectrometer (MS) with real time gas analysis capability, is developed. The main goal of performing the experimental study is to investigate the physical and chemical processes that take place when reactive flow goes through the reactor. This includes fuel conversion, oxidation and reduction reactions of the OC, OC material selection and optimization, and the temperature variation and distribution along the reactor channel. Results of the experiments will be analyzed and used to validate the numerical simulations.

---

[i] Z.Zhao, T.Chen, A. F. Ghoniem. Rotary bed reactor for CLC with carbon capture. Part I: Reactor design and model development. Energy Fuels, 2013, 27 (1), pp 327–343

[ii] Z.Zhao, T.Chen, A. F. Ghoniem. Rotary bed reactor for CLC with carbon capture. Part II: Base case and sensitivity analysis. Energy Fuels, 2013, 27 (1), pp 344–359

(816e) Comparison of Various Carbon Capture Technologies to Reduce CO2 Emissions From a Cement Plant

Ozcan, D. C., University of Edinburgh
Bocciardo, D., University of Edinburgh
Ferrari, M. C., University of Edinburgh
Ahn, H., University of Edinburgh
Kierzkowska, A. M., ETH Zurich
MÃ¼ller, C. R., ETH Zurich
Brandani, S., University of Edinburgh
Originally presented on: 11/8/2013 13:50:00 - 14:10:00

Substantial reductions of global CO_2 emissions by at least 50% are required by 2050 according to Intergovernmental Panel on Climate Change [1]. While CO_2 reduction technologies have been mostly developed for the purpose of carbon capture from the power generation industry, it is important to extend this to the cement industry in order to achieve the CO_2 capture target projected over the next 40 years. The cement industry accounts for over 2 billion tonnes of CO_2 emitted per year [2,3]. Only marginal reductions in CO_2 emissions can be achieved through improving the energy efficiency of the cement manufacture process. Significant reductions can be obtained only if a carbon capture unit is integrated in a cement plant given that around 60% of CO_2 emission comes from the calcination of the raw material.

Both post-combustion and oxy-firing would be deemed appropriate for carbon capture from cement works. To compare the different configurations full process flowsheets have been developed and simulated in Unisim Design Suite from Honeywell. The process considered is based on the reference case from the IEA [4] and includes the main raw material decomposition and clinkerization reactions. The base case simulation was then modified to integrate three different post-combustion capture processes: amine-scrubbing; calcium looping process; and membrane-based capture units. Furthermore, an advanced configuration of Ca-looping process has been investigated where the energy intensive air separation unit was replaced with a chemical looping combustion (CLC) cycle. Each case has been optimised to minimise its energy consumption and compared in terms of CAPEX/OPEX and its resulting CO_2 avoided cost at the same capture rate.

[1] Intergovernmental Panel on Climate Change (IPCC) (2007b). Metz, Bert. Davidson, Ogunlade. Bosch, Peter. Dave, Rutu. Meyer, Leo (Eds.). *Climate Change 2007: Mitigation. Contribution of Working Group III to the Fourth Assessment Report of the Intergovernmental Panel on Climate Change.* Cambridge University Press, Cambridge, United Kingdom and New York, NY, USA.

[2] CW Group *Global Cement Volume Forecast Report*, 2012.

[3] IEA Carbon Emission Reductions up to 2050. In *Cement Technology Roadmap 2009*; International Energy Agency: Paris, 2009.

[4] IEA *CO_2 Capture in the Cement Industry*, July 2008/3, 2008.

(816f) Non-Isothermal Performance Modeling for CO2 Adsorption

Breault, R. W., National Energy Technology Laboratory
Huckaby, E. D., National Energy technology Laboratory
Shadle, L. J., National Energy Technology Laboratory
Spenik, J., REM, Corporation
Originally presented on: 11/8/2013 14:10:00 - 14:30:00

The National Energy Technology Laboratory as well as other institutions are in the process of conceptualizing new technology as an alternative to amine based absorption processes for CO2 capture from large sources such as utility power generation facilities. Many, but not all of these advanced dry processes are based upon sorbents composed of supported polyamines. This particular investigation gives a comparison between the predicted performance with a non-isothermal model and the experimental performance of one of these new supported polyamine sorbents. In the continuous process CO2 from is absorbed from flue gas in the absorber, a cyclone separates the carbonated particles from the de-carbonated flue gas and then regenerates the sorbent, creating a concentrated stream of pure CO2 for sequestration. In the current work, the reactors are run in a batch mode, that is without circulating solids. In this work, experimental data from adsorption and regeneration in a fluidized bed operated in batch mode are compared to simulation predictions from a 3-dimentional multiphase flow model complete with chemistry and thermal effects. The effects of gas flow rate and reactor temperature are explored and compared with experimental measurements. The specifics for the experimental facility and the model as well as the comparative analysis between the data and the simulation results will be discussed.

(816g) A Correlation for the Impact of Extracting Steam From Power Plants for Carbon Capture Processes

Berger, A. H., Electric Power Research Institute
Bhown, A. S., Electric Power Research Institute
Originally presented on: 11/8/2013 14:30:00 - 14:50:00

A Correlation for the Impact of Extracting Steam from Power Plants for Carbon Capture Processes

Adam H Berger and Abhoyjit S Bhown

Electric Power Research Institute (EPRI), Palo Alto, CA

Abstract

Post combustion carbon dioxide capture and storage (CCS) from power plants is an energy intensive process often consuming one quarter to one third of the total power generated by the base power plant. This parasitic energy is comprised of two main terms: the auxiliary load, which accounts for pumping, compression, and other mechanical or electrical work; and the thermal load, which is typically steam extraction for process heating. It is important to understand how these two terms interact with the base power plant so that accurate parasitic energy can be calculated and CCS materials or processes can be developed and compared based on actual impact on the power output of power plants.

While the impact of the auxiliary load on the power plant output is immediately apparent, the impact of the thermal load on the power plant is more difficult to quantify. The thermal load is used to provide heating energy to a CCS process in the form of steam extracted from the steam cycle of the power plant. The extracted steam is used to provide process heating such as in the reboiler of a solvent stripper or heating solid material in a temperature swing adsorption process. The electrical impact of extracting steam is defined as the power loss in the turbines from the decreased steam flow rates. This power loss is a function of the temperature and pressure of steam extraction as well as the quantity of steam extracted. The heating energy has components, the energy from de-superheating the steam extracted and energy from the heat of vaporization of steam condensing at the pressure of extraction. This condensation at the saturation temperature also determines the temperature at which process heating can be provided. In addition to the temperature and pressure of the steam extracted, the quantity of steam extracted is also important because the effect of steam extraction on low-pressure turbine operation and efficiency.

In this work, we use state-of-the-art models of low pressure turbines in power plant steam cycles to quantify the effect of steam extraction on power plant operation. We further develop a correlation for the impact on a power plant of providing thermal energy at a specified quantity and temperature to a CCS process. This correlation can be used in any process or material development to more accurately compare the effect of different process configurations or material properties on the total energy impact of CCS and thus compare trade-offs between mechanical work and thermal load or between capital and operating expenses.

AUTHOR INDEX

AUTHOR INDEX

AUTHOR INDEX

AUTHOR INDEX